*RECREATIONAL
FLYING*

ALSO BY RICHARD L. TAYLOR

Instrument Flying (Revised)

Fair-Weather Flying (Revised)

Understanding Flying

Positive Flying (with William M. Guinther)

IFR for VFR Pilots

The First Flight

RECREATIONAL FLYING

The Complete Guide to Earning and Enjoying the NEW Recreational Pilot Certificate

Foreword by Jack J. Eggspuehler, *President, National Association of Flight Instructors*

RICHARD L. TAYLOR

AN ELEANOR FRIEDE BOOK

Macmillan Publishing Company · New York

Collier Macmillan Publishers · London

Copyright © 1990 by Flight Information, Inc.

Illustrations on pages 50, 53, 66, 114, 121, 171, 174, 177, 178, 179, 183, 191, 195, 197, 204, 207, and 208 are by Paul C. Haynie and are from *Understanding Flying* by Richard L. Taylor

All rights reserved. No part of this book may be reproduced or transmitted in any form or by any means, electronic or mechanical, including photocopying, recording or by any information storage and retrieval system, without permission in writing from the Publisher.

Macmillan Publishing Company
866 Third Avenue, New York, NY 10022
Collier Macmillan Canada, Inc.

Library of Congress Cataloging-in-Publication Data
Taylor, Richard L.
   Recreational flying : the complete guide to earning and enjoying the new recreational pilot certificate / Richard L. Taylor.
      p. cm.
   "An Eleanor Friede book".
   ISBN 0-02-616635-6
    1. Airplanes—Piloting. 2. Air pilots—Licenses—United States.
  I. Title.
TL710.T3625   1990
629.132'52—dc20   89-12850
                                                                       CIP

Macmillan books are available at special discounts for bulk purchases for sales promotions, premiums, fund-raising, or educational use. For details, contact:

Special Sales Director
Macmillan Publishing Company
866 Third Avenue
New York, NY 10022

10 9 8 7 6 5 4 3 2 1

Designed by Jack Meserole

PRINTED IN THE UNITED STATES OF AMERICA

To those pilots who take inspiration
from this book and use it as a
springboard to an aviation career or
a lifetime of flying just for fun

CONTENTS

FOREWORD by Jack J. Eggspuehler, *President, National Association of Flight Instructors*   ix

Introduction   xiii

Definitions and Descriptions: A Glossary of "Pilot Talk"   xvii

1   Flying Guidelines: The Rules and Regulations   3

2   Airplane Structures and Components   21

3   Basic Aerodynamics   31

4   Basic Flight and Engine Instruments   48

5   Basic Flight Operations   57

6   Airplane Engines: Principles and Operation   70

7   Preflight and Engine Start   78

Contents

8	Ground Operations: Taxiing and Before-Takeoff Procedures	84
9	Training Operations	92
10	Normal Takeoffs and Landings	109
11	Abnormal Landings	123
12	Crosswind Takeoffs and Landings	134
13	Airport Operations	145
14	Airspace Considerations	153
15	Basic Aerial Navigation	160
16	Meteorology for the Recreational Pilot	168
17	Aircraft Loading	190
18	Aircraft Performance	200
19	Maximum Performance Operations	211
20	Emergency Procedures	219
21	Safety of Flight Procedures	223
	Index	229

FOREWORD

Dick Taylor's books don't need introductions. As one who saw the need for improved training in aviation at all levels, he has continually invested himself in his writing to make flying safer as well as more fun. This book is no exception.

Several years ago, the Federal Aviation Administration (FAA), in response to the recommendation of an industry committee, adopted some revised standards for pilot certification. There were new demands on pilots operating into high-density airports as the airspace became far more complex. Among those standards were increased communications and navigation skills, integrated instrument instruction, and significantly higher standards for the operation of aircraft in complex airspace. In some ways, this timely new emphasis was necessary to operate in our changing environment.

But there was also a growing number of pilots who didn't want to develop the skills necessary to fly cross-country, or around busy airports. Such pilots chose to operate their airplanes close to home, in much less congested airspace, using the airplane as a recreational vehicle instead of a transportation tool. Recognizing this need, a blue-ribbon committee was formed in 1981 to make recommendations for another revision of pilot certification standards.

Foreword

As a result of this committee's work, the FAA was petitioned in 1982 to consider the concept of a Recreational Pilot. The thrust of this petition was the notion that it is time to consider breaking down private pilot training into two possible steps. The first step would consist of teaching the aspiring pilot the basics of piloting, with a high level of skill required in flying the airplane. At that level, the Recreational Pilot would be capable of carrying a passenger, but would be limited in the scope of operations he could undertake. For instance, the pilot would be limited to certain geographical areas and the altitudes to which he could fly would be restricted. As a result, far greater emphasis would be placed on the fundamental skills which make for safer flying.

In this book, Dick Taylor brings you back to those basics. He manages to explain the fundamentals of flying in an understandable fashion so that you can appreciate not only the "how," but also the "why." It is his explanation of the application of skills to real-world situations that makes Dick's book so important to the future pilots of our country.

Make no mistake: the Recreational Pilot described in these pages is *not* a lesser pilot. Quite the contrary. Even though not very experienced, he is very capable of handling the airplane. Once the airplane is mastered, the newly certificated Recreational Pilot may take friends or family for rides while gaining additional experience—qualified and capable of flying locally in every way, but not quite ready to go outside his home territory.

Once he nails down the Recreational Pilot skills, there is no reason for this pilot to stop. The knowledge and skills required can be further honed as he proceeds to higher levels of capability—the Private Pilot certificate and beyond. With the Private Pilot certificate, he can use the airplane to go places and enjoy the airplane for transportation.

Foreword

The procedures and techniques set forth in this book are not just things that are "nice to know" or "nice to do"; rather, they are the basic skills of flying that a pilot can use to safely fly recreationally for a lifetime. On the other hand, they represent a foundation that he can use for a lifetime of safe Private Pilot flying, or for a professional career in corporate or airline operations.

When it comes to flying, almost everyone aspires to do it right. With this book you can come to know how to do it right. And with that kind of a start, you can build a flying foundation which will last a lifetime.

Jack J. Eggspuehler
President,
National Association of
Flight Instructors

INTRODUCTION

The addition of the Recreational Pilot Certificate to the list of credentials available to U.S. pilots is an important step in the continuing development of general aviation. It provides access to personal aviation at an affordable level, it gives you a chance to find out if you really *like* flying without committing a large sum of money and a lot of your time, and it is the foundation on which you can build more advanced piloting skills . . . or, if you like, you can stay at the Recreational Pilot level and enjoy its privileges for many years to come.

Regardless of the level to which you aspire, aviation requires more of its participants than almost any other part-time activity, for without knowledge and minimum skills, even the smallest, least-complicated airplane can put you into a hazardous situation in a hurry. There's an old saying in this business (after a while, you'll discover that there are a *lot* of old sayings in this business!) to the effect that "aviation is not inherently dangerous, but like the sea, it is terribly unforgiving." That's very true—and knowledge will be a big part of your safe and pleasant career as an airplane pilot. *Recreational Flying* provides a solid base of knowledge in that respect, but you must be willing to benefit from each actual experience, to learn something from every flight, and above all, you must believe that no one "knows it all" about

Introduction

flying. The objective of this book is to acquaint you with basic aeronautical knowledge, procedures, and piloting techniques; your Flight Instructor will elaborate as your training progresses.

Recreational Flying is based on regulations and procedures currently in effect, but you need to understand that some of these can—and no doubt will—change with time. *It's up to each pilot to maintain his knowledge of regulations and procedures at a current level.*

In addition to the "book learning" contained in these pages, *Recreational Flying* suggests some pilot techniques for the various operations required of you. Sooner or later, you'll discover that your Flight Instructor will want you to do something differently; don't be upset, because there are probably as many different instructional techniques as there are instructors. Rest assured that the suggestions in this book are safe and proper, and are founded on the techniques and procedures on which the minimum checkride standards are based.

Recreational Flying was written with standard, factory-built airplanes in mind—the stock airplanes which are most likely to be used by pilots at this level of certification. But on occasion, some Recreational Pilots may get involved with nonstandard airplanes—home-builts, for example. In this situation, be very certain that whatever flight operations you intend to accomplish lie well within the operational limits of the airplane at hand. You're the responsible party, so when in doubt, ask; and if you don't get a good answer, *don't try it!* The bottom line in this regard is the Pilot's Operating Handbook (or equivalent document); the limitations, instructions, and procedures found there are the final authority.

Recreational Flying has been written on a rather limited

Introduction

scale with regard to the entire body of aviation knowledge, since it is intended to serve only those who aspire to the Recreational Pilot Certificate. Therefore, you'll not find explanations of principles, regulations, or procedures which are required for more advanced pilot certificates. There's nothing here about night flying, radio communications, simulated instrument operations, or flying into or out of controlled airports. As your ambitions and objectives expand beyond the Recreational Pilot Certificate, it will be necessary to consult the appropriate materials—there is always more to learn!

You will need to study a rather large body of material in order to prepare for the Recreational Pilot written examination. Your Flight Instructor will guide you through this part of the training, or will refer you to someone who specializes in ground instruction. In either case, you can help yourself considerably by augmenting this book with other sources of information, and I strongly recommend that you obtain copies of three government publications: the *Airman's Information Manual*, the *Pilot's Handbook of Aeronautical Knowledge* (Advisory Circular No. 61-23B), and the *Flight Training Handbook* (Advisory Circular No. 61-21A). The first of these is a quarterly publication which contains the basic information and procedures required to fly in the national airspace system; you may want to maintain your own subscription, or perhaps you can find a recent copy somewhere around the airport. The other two should be part of your pilot's library; the information they contain is relatively timeless, and will be a big help as you move through your training and continue to develop your skills as a pilot.

So, from a rich background of nearly thirty-five years of flying a wide variety of aircraft, twenty-two of those years

Introduction

spent in aviation education at Ohio State University, I offer you my thoughts on what you need to know to become a safe, proficient Recreational Pilot. May this new certificate be your gateway to pleasant and rewarding experiences.

Good Luck!
RICHARD L. TAYLOR

DEFINITIONS AND DESCRIPTIONS: A GLOSSARY OF "PILOT TALK"

As you begin your study of the fundamentals of flying, you will discover that aviation people have a language all their own. And even after you have acquired some experience as an aviator, you'll come across terms and acronyms and mnemonics that are widely used but narrowly understood.

With that in mind, this rather comprehensive glossary of definitions and descriptions will be helpful when you come up against a concept that needs more illumination, or a bit of jargon that no one seems to understand. Take a few minutes to scan the glossary before you begin reading the book proper, to give you an idea of its contents. You might even want to place a paper clip or bookmark so that you can refer to the glossary quickly as you move through the pages that follow.

abeam The position of an object (an airport, a town, another airplane) directly off either wingtip of an airplane.
accelerated stall A training exercise in which the airplane is subjected to acceleration (G loading), then stalled, to demonstrate

Definitions and Descriptions: A Glossary of "Pilot Talk"

that a stall will occur at a higher-than-normal airspeed in a turn.

acceleration error On headings of east or west, a magnetic compass will erroneously indicate a turn toward the north when the airplane accelerates, even though the airplane's heading changes not at all. (*See also* deceleration error.)

active runway The runway currently in use, identified by the first two digits of its magnetic orientation. Commonly shortened to "the active."

advection fog Fog which forms as the result of warm, moist air moving over a colder surface. Frequently found in coastal areas when air flows from the water onto the shore.

adverse yaw The tendency of almost all airplanes to swing the nose in the opposite direction when a turn is begun. Adverse yaw is prevented or corrected by proper use of the rudder.

aerodynamic A catch-all term which refers to the forces generated by, or the results of, air ("aero") moving ("dynamic") over or around the various surfaces of an airplane.

aeronautical charts The term applied to a flyer's "road maps." These are maps specifically designed to provide the information a pilot needs to navigate from point to point.

AGL (Above Ground Level) An altitude term used to indicate height of cloud layers and obstructions, to define certain types of airspace, and as a reference for an airplane's actual height above the surface.

aileron One of the primary flight controls, ailerons are located near the outboard trailing edges of the wings, and are primarily responsible for roll control.

AIM (Airman's Information Manual) An FAA publication which contains procedures, airport listings, facilities, and just about everything about aviation operations that isn't published anywhere else.

airfoil The shape of any surface on an airplane which generates the aerodynamic force of lift whenever air is moved over or around the surface.

Definitions and Descriptions: A Glossary of "Pilot Talk"

air mass A volume of atmosphere which can be identified and classified with respect to its temperature, moisture content, and pressure.

AIRMET A weather advisory about conditions such as low visibilities or high winds that will affect primarily light-plane operations.

Airport Traffic Area A cylinder of airspace five miles in radius and 3,000 feet deep over an airport with an operating control tower. This airspace is *off limits* to Recreational Pilots!

airspeed *See* various types, such as indicated, true.

airspeed indicator An instrument that measures the difference between static and dynamic air pressure; it indicates the velocity of the airplane through the air.

Airworthiness Directive (AD) A bulletin from the FAA to indicate something's wrong with a particular make and model of airplane, or some appliance installed thereon. Unless ADs are complied with, the affected airplanes are not considered airworthy.

Alert Area A part of the airspace in which aviation activities are not officially restricted, but in which extra vigilance should be maintained because of a high volume of air traffic or some unusual type of flying, or both.

altimeter An instrument which measures air pressure and displays that measurement to the pilot in terms of height or altitude.

altitude *See* specific types, such as true, pressure, density.

angle of attack The angle (in degrees) at which an airplane's wing meets the oncoming air. Also described as the angle between the relative wind and the chord line of the wing.

angle of incidence The angle (in degrees) at which the wing is mounted on the airplane, relative to the airplane's nose-to-tail centerline.

A&P mechanic A technician specifically trained and then certificated by the FAA to perform repairs on Airframes and Powerplants.

Definitions and Descriptions: A Glossary of "Pilot Talk"

approach/landing stall A training operation in which the airplane is stalled in a configuration and condition of flight similar to that used during an approach to land.

ASL (Above Sea Level) Means the same as mean sea level (*see*).

atmosphere The relatively shallow blanket of air surrounding the earth.

avgas Contraction for "aviation gasoline."

avionics Contraction for "aviation electronics," referring to all types of electronic equipment aboard an airplane.

azimuth A position measured in angular degrees. Used in aviation to describe the relation of an aircraft to some other object, such as a geographic feature, radio station, or another airplane.

bank Movement of an airplane in which one wing drops, the other rises. "Bank" is synonymous with "roll," and is primarily controlled with the ailerons.

bearing Position of an object or radio station stated in terms of azimuth. Can be either to or from the object or the airplane.

Bernoulli principle The physical "law" which states that whenever the flow of air is speeded up, its pressure decreases—the basic principle of lift production.

best angle of climb The result of climbing an airplane at a specified airspeed which produces the greatest gain in altitude over a given horizontal distance. Useful for clearing nearby obstacles in the flight path.

best rate of climb The result of climbing an airplane at a specified airspeed which produces the greatest gain in altitude in a given period of time. Most useful for "normal" climbs, when it is important to achieve cruising altitude in the shortest possible time.

BFR (Biennial Flight Review) The every-other-year visit to the FAA or a Certificated Flight Instructor to make sure you're not practicing your mistakes. You may not act as pilot-in-command unless you have had a documented, satisfactory BFR within the

Definitions and Descriptions: A Glossary of "Pilot Talk"

previous two years. *For non-instrumented-rated pilots who have logged less than 400 hours of flight time, this is an annual requirement.*

buffet   The shaking and rumbling of an airplane caused by the turbulent flow of air over and around its surfaces when a stall is imminent.

calibrated airspeed (CAS)   Indicated airspeed corrected for installation and position errors. All performance and limiting airspeeds are quoted in terms of CAS.

camber   The curvature of the upper surface of an airplane's wings and other aerodynamic surfaces, which causes airflow to accelerate, thereby producing lift.

carbon monoxide   One of the chemical by-products of the burning of gasoline in an airplane engine. If this odorless, colorless, tasteless gas gets into the cabin and is inhaled by the pilot, incapacitation may occur.

carburetor   The device on an airplane engine that mixes gasoline vapor and air so that it can be burned and produce power.

carburetor heat   Refers to the introduction of heated air into the carburetor to prevent ice formation or to melt away ice that may have already formed.

carburetor ice   That which forms in the throat, or main air passage, of an airplane's carburetor. In sufficient quantity, carburetor ice can literally "choke" the engine and cause it to quit running.

cardinal headings   The eight directions most generally referred to in orientation or navigation; north, east, south, west, and northeast, northwest, southeast, southwest.

category (of aircraft)   With respect to the privileges and limitations imposed on pilots, aircraft are placed into broad categories: airplane, helicopter, glider, balloon, for example.

ceiling   The lowest cloud layer that covers enough of the sky to be classified "broken" or "overcast." If the observer calls either of

Definitions and Descriptions: A Glossary of "Pilot Talk"

these layers "thin," it is not considered a ceiling. A Recreational Pilot may *never* fly through a ceiling, either going up or coming down.

Celsius An expression of temperature in degrees; replaces the term "centigrade." In aviation, all temperatures other than those at the surface are given in degrees Celsius, or simply degrees "C."

centrifugal force When a body moves in a circular path, centrifugal force exerts a pressure toward the outside of the circle. You feel it on thrill rides at the amusement park, and you'll feel it (to a lesser degree) whenever your airplane rolls into a turn. The steeper the bank, the greater the centrifugal force.

CFI A Certificated Flight Instructor.

CG (Center of Gravity) The apparent balance point of an airplane. Manufacturers specify an allowable range through which the CG may move; it's up to the pilot to be certain that the airplane is always loaded properly with respect to its CG limits.

checklist A list of items or operations to be checked before takeoff or landing or other critical phases of flight.

checkpoint A navigational fix; a recognizable landmark used as a "point" to "check" one's progress during a cross-country flight. (*See* cross-country.)

checkride An evaluation flight, conducted by either an FAA inspector or a designated pilot examiner, for the purpose of determining your competence for a pilot certificate or rating.

chord The imaginary line, and the distance, between the leading and trailing edges of an airfoil.

CIFFTRS A "memory-jogger" checklist (pronounced "sifters") to be used before takeoff to make sure that the essential things have been taken care of. Each letter stands for a word: Controls, Instruments, Fuel, Flaps, Trim, Runup, Seat belts.

circuit breaker A device which breaks an electrical circuit whenever a potentially damaging amount of current is sensed. Much like a fuse, except that a circuit breaker can usually be reset by the pilot when the problem is resolved.

Definitions and Descriptions: A Glossary of "Pilot Talk"

cirrus clouds High (typically 20,000 feet or more), thin, wispy, clouds composed of ice crystals.

closed runway For any of a number of reasons, an airport operator may choose to make a runway unavailable to pilots for takeoff or landing. The universal marking in this situation is a large yellow or white X placed on each end of the affected runway.

closed traffic That flight operation in which a pilot flies his airplane in a rectangular pattern around the airport for the purpose of practicing repetitive takeoffs and landings.

cold front The leading edge of a moving mass of cold or cool air.

compass A device that indicates direction by sensing the earth's magnetic field.

compass correction card A card placed in every airplane to indicate installation errors in the compass reading on various cardinal headings.

compass deviation The errors inherent in a specific compass installation due to spurious magnetic influences within the airplane itself.

compass heading The direction an airplane is pointed, based on the indication of its magnetic compass.

compass rose A series of radial lines painted on the airport surface (usually on a convenient taxiway or ramp) so that an airplane can be lined up on the cardinal headings for calibration of the magnetic compass.

controllable-pitch prop A propeller installation in which the blades can twist in response to loading or pilot input; by changing pitch (twisting), the propeller changes its speed of rotation and thrust production.

controlled airport Any airport with an operating control tower, and one of the facilities not available to Recreational Pilots.

controlled airspace A catch-all term which refers to any airspace in which ATC controllers have the responsibility for traffic separation when weather conditions are less than basic VFR minimums.

Definitions and Descriptions: A Glossary of "Pilot Talk"

convection A weatherman's term applied to any situation in which air is being lifted because of surface heating. Usually implies cumulus development in any clouds that may form as a result.

convective SIGMET A weather advisory which warns of potential or existing thunderstorm development in a specific geographical area. A convective SIGMET should command the attention of *all* pilots, and usually creates a no-go situation for the Recreational Pilot.

coordinated turn A turn in which the angle of bank is exactly that required for the rate of turn, with the airplane neither slipping nor skidding. A coordinated *entry* implies that rudder and ailerons are used properly to prevent adverse yaw.

coordinates The numerical values of latitude and longitude which locate any point on the surface of the earth.

course Same as "track," or the imaginary line across the surface that describes the actual flight path made good by your airplane.

cowl flaps Pilot-controlled doors or shutters installed on some higher-powered airplanes to facilitate engine cooling; cowl flaps permit adjustment of the amount of cooling air passing through the engine compartment.

cowling The sheet metal (or fiberglass) that covers the engine of an airplane, making it more streamlined and also facilitating the flow of cooling air.

crab angle The number of degrees a pilot must turn his airplane into the wind in order to maintain a straight track across the ground. When this is being accomplished, the airplane is actually moving somewhat sideways, or "crabbing."

critical angle of attack The angle (in degrees) between relative wind and the chord line of the wing which, when exceeded, produces an aerodynamic stall.

cross-country A flight that takes off from one airport and lands at another more than 25 nautical miles away.

Definitions and Descriptions: A Glossary of "Pilot Talk"

crosswind The flow of air at an angle across a runway, or across an intended path of flight. In either case, the pilot must make corrections in order to fly in a straight track across the ground.

crosswind leg *See* traffic pattern.

cruise A general term applied to that phase of flight in which an airplane has reached the desired altitude and airspeed for a trip; the "level" portion of a flight.

cruising altitude The height at which you are flying or intend to fly on a particular trip or segment thereof.

cumulus clouds "Heaped up," billowy clouds which exhibit vertical development, and which sometimes grow into thunderstorms.

datum The point from which measurements are made when calculating the weight and balance of an airplane.

dead reckoning A method of navigation in which a pilot estimates his arrival over the next checkpoint, that estimate based on his knowledge of the time over the previous checkpoint, and calculations for wind effect, airspeed, and heading. "Dead" is actually a corrupted abbreviation of the word "deduced."

deceleration error On headings of east or west, a magnetic compass will erroneously indicate a turn toward the south when the airplane decelerates, even though the airplane's heading changes not at all. (*See also* acceleration error.)

density altitude Pressure altitude corrected for nonstandard temperature. The higher the density altitude, the lower an airplane's performance.

Designated Pilot Examiner A highly qualified and experienced flight instructor who has undergone special training in pilot evaluation, and who has been empowered by the FAA to give checkrides and award pilot certificates and ratings.

dihedral A design feature which improves the roll stability of an airplane. When dihedral is part of the design, the wingtips will be higher than the wing roots. Most low-wing airplanes have

Definitions and Descriptions: A Glossary of "Pilot Talk"

considerable dihedral; most high-wing airplanes have little or none.

displaced threshold When certain conditions exist, it may be unwise (or perhaps even unsafe!) to attempt a landing at the very beginning of the paved portion of the runway. Hence a "displaced threshold"—a painted line which indicates the point where the runway actually begins for the purpose of landing.

drag That aerodynamic force which acts to hold an airplane back. It is usually overcome by the force of thrust. There are several kinds of aerodynamic drag, but it's the sum total that is important to the pilot.

dual instruction Periods of flight training during which a flight instructor occupies one of the pilot seats in the airplane. Commonly shortened to "dual" (as opposed to "solo").

ear block The difficulty occasionally encountered when descending rapidly (you've no doubt experienced this during a fast elevator ride). The solutions are to swallow or to work your jaws in order to facilitate the equalization of pressure between the inner and outer ear passages.

empty weight The weight of the airframe itself, with no passengers, no cargo, and only unusable fuel and undrainable oil.

FAA The Federal Aviation Administration.

FARs The Federal Aviation Regulations.

FBO (Fixed-Base Operator) The organization that sells gasoline, airplanes, and flight services at an airport.

field elevation The official height of an airport expressed in feet above mean sea level (MSL). Normally used to set the altimeter when there is no official altimeter setting available.

fin The fixed vertical surface that is part of the empennage on most airplanes.

final approach *See* traffic pattern.

fixed gear Term applied to an airplane with nonretractable wheels.

Definitions and Descriptions: A Glossary of "Pilot Talk"

fixed-pitch prop A propeller made of one solid piece of metal or wood, whose blades are unable to twist or change pitch. This type of propeller is usually found on lower-powered airplanes.

flaps Devices that extend from the trailing edge of an airplane's wings in order to increase drag and allow the airplane to be landed at a lower speed, and with a steeper final approach path.

flare The change in pitch attitude in the final stage of a landing, for the purpose of reducing airspeed and rate of descent. Also known as "roundout."

flight check The flight portion of a pilot evaluation for a certificate or rating. Same as "checkride."

flight path The actual track of an airplane across the ground.

flight plan A pilot's intended course of action; may be formally filed with the Air Traffic Control System, or may simply be stored in the pilot's mind.

Flight Service Station (FSS) A facility of the FAA set up to provide virtually any operational service to aviators, including weather information, communications, and emergency service.

flight time That time an airplane spends in flight, or for pilot logbook purposes, the total time from leaving the ramp until returning to the ramp after a flight.

fog Stratus (layered) clouds which form very close to the surface. Fog is always the result of moist air being cooled to its saturation point.

forced landing Any unscheduled landing, whether occasioned by mechanical failure, fuel exhaustion, or weather.

four-cycle engine An internal combustion engine that incorporates four operating strokes of the piston; intake, compression, power, and exhaust.

front The boundary, or zone of dissimilarity, between two different types of air masses.

FSDO (Flight Standards District Office) The general aviation pilot's primary contact with the FAA.

fuel drain A fitting that is located in a low point in the fuel system

Definitions and Descriptions: A Glossary of "Pilot Talk"

for the purpose of draining accumulated water and other impurities from the fuel supply.

fuel pump A pump (either engine-driven or electrically powered) intended to guarantee a supply of fuel under pressure to the engine.

fuel selector valve On airplanes with multiple tank arrangements, this valve permits selection of one or more tanks to supply the engine with fuel.

fuel strainer A very fine metal screen capable of blocking the passage of impurities in fuel, thereby protecting the engine and assuring a clean fuel supply.

fuel vent An opening from the fuel tank to outside air pressure so that a vacuum is not created inside the tank as fuel is used.

fuse A device that breaks an electrical circuit when a potentially damaging amount of current is sensed. When a fuse "blows," it must be replaced before current can be restored.

fuselage The main body of an airplane, with space for passengers and cargo, and which also serves as the support structure for wings, tail, engine, landing gear.

general aviation The term applied to all aviation activity in the U.S. except the military and scheduled airlines.

G force The force of gravity, which is exerted on all objects on the surface of the earth. As an airplane turns, or pulls out of a dive, G force increases and is manifested by a feeling of heaviness, or increased weight.

glide path The vertical flight path described by an airplane on its approach to land.

GMT (Greenwich Mean Time) The standard time used by aviators throughout the world. GMT represents the time in Greenwich, England. (*See also* Zulu.)

gross weight The total weight of an airplane as loaded for takeoff. Every airplane is limited to a *maximum* gross weight.

ground effect The increase in an airplane's performance when

Definitions and Descriptions: A Glossary of "Pilot Talk"

within about one-half wingspan of the ground. It causes airplanes to float on landing, and sometimes causes pilots to think they're safely airborne, only to find out that the airplane can't climb when *out* of ground effect.

ground fog Stratus clouds formed very close to the surface; this term is used primarily to indicate that it's not sea fog.

ground loop A sudden swerve of 180 degrees, usually occurring at the end of the landing roll in a poorly controlled tailwheel airplane. Can sometimes be useful as a last resort when a landing roll turns out to be dangerously long ... such as if you find yourself running into trees, or a ditch, or another airplane.

ground reference maneuvers Pilot-training exercises designed to develop the ability to fly a predetermined track across the ground, such as S-turns across a road, turns about a point, and rectangular patterns.

ground roll The distance on the ground required to accelerate an airplane to takeoff speed, or the number of feet required to bring the airplane to a stop after touchdown.

groundspeed The speed made good across the ground; true airspeed minus a headwind, plus a tailwind. Groundspeed is the heart of accurate aerial navigation.

ground trainer Any one of a number of devices used to simulate flight operations. A ground trainer might be a simple desktop unit, or a full-fledged simulator with motion, visuals, and almost total realism.

GUMPS The hand-in-glove companion of CIFFTRS. Again, each letter stands for a word reminding a pilot of essential items to be checked prior to landing: Gas, Undercarriage (landing gear), Mixture, Prop (if controllable-pitch), and Seat belts.

heading The direction an airplane is pointed; expressed in angular degrees measured from magnetic north. Same as "course" or "track" when there is no wind correction.

headwind Wind blowing from any point 90 degrees either side of

Definitions and Descriptions: A Glossary of "Pilot Talk"

an airplane's course. Headwind is always subtracted from true airspeed to obtain groundspeed.

heel brakes Small brake pedals on the floor immediately beneath the rudder pedals in many older airplanes.

hold line A line painted across a taxiway to indicate the point behind which a departing airplane should hold until ready to take off.

horizontal stabilizer The fixed horizontal portion of the empennage, intended to stabilize the pitch attitude of the airplane.

humidity An expression of the amount of moisture in the air. *Relative humidity* is actual moisture content compared to the amount of moisture a parcel of air can hold as vapor, expressed as a percentage—100 percent relative humidity indicates completely saturated air. Airplane performance is always decreased somewhat as humidity increases.

idle The closed-throttle condition of an airplane engine. On the ground, most engines will idle at 500 to 700 RPM; in the air, idle speed will depend on airspeed.

IFR (Instrument Flight Rules) The regulations covering operations under less-than-visual weather conditions; also a general term applied to all instrument operations.

ignition system The magnetos, spark plugs, switches, and associated wiring that provide the electrical energy inside the engine cylinders to ignite the fuel-air mixture and produce power.

indicated airspeed The number of knots (or miles per hour) at the end of the pointer on the airspeed indicator. Subject to several errors, but reliable enough for nearly all flight operations in low-speed airplanes.

Inspector The title of a person employed by the Federal Aviation Administration to help oversee the operations of a particular segment of aviation. Inspectors process complaints, investigate accidents and alleged violations of the FARs, and, on occasion, conduct pilot checkrides.

Definitions and Descriptions: A Glossary of "Pilot Talk"

intersection The crossing of two runways, or of a runway by a taxiway.

intersection takeoff A practice used by some pilots when there is sufficient runway remaining between an intersection and the end of the runway; departing from the intersection often saves considerable time and taxi distance. *To be used only when there's absolutely no concern about the safety of using less than full runway length.*

knot One nautical mile per hour.

landing The act of returning an airplane to the ground. There are several types of landings: normal, soft-field, short-field, three-point, and more. (*See* various types.)

landing direction indicator A means of indicating wind direction, and therefore the recommended landing direction. The usual indicators are wind socks or wind tees, but flags, plumes of smoke, or wind patterns on bodies of water may also be used.

landing distance The number of feet required for the entire landing procedure, usually from a fixed point, such as an altitude of 50 feet above the runway threshold. Landing distance includes the air distance necessary to lose that 50 feet, plus the ground distance after touchdown.

landing roll The number of feet required to bring an airplane to a stop after touchdown.

lateral axis The imaginary line that runs from wingtip to wingtip; movement about the lateral axis is called "pitch."

latitude Geographical measurement on the earth's surface, beginning at the equator and proceeding north and south. In conjunction with longitude, locates any point on the earth's surface. Always expressed in degrees and minutes (60 minutes to each degree).

leaning The process of adjusting the fuel-air mixture being used by an airplane engine in order to promote smoother engine operation and fuel efficiency.

Definitions and Descriptions: A Glossary of "Pilot Talk"

load factor   The number of Gs being imposed on an airplane structure at any given time; when the G force is 2, the load factor is also 2. Most general aviation airplanes are designed for a load factor of 3.8, or 3.8 times the force of gravity. Most humans get uncomfortable when subjected to a load factor of 2.

loading chart   A chart used in calculating and determining the safe loading of an airplane. There are several types in use, but all of them arrive at a comparison of the airplane's weight and the distribution of that weight.

logbook   A pilot's record of his aerial activities. The law requires that pilots maintain a log for the purpose of establishing currency and recency of flight experience.

longitude   A geographical measurement on the earth's surface, beginning at the prime meridian (which runs through Greenwich, England), and proceeding east and west to the other side of the world. In conjunction with latitude, enables the locating of any point in the world.

longitudinal axis   An imaginary line which runs from nose to tail of an airplane. Movement about the longitudinal axis is called "roll."

mag check   Jargon for "magneto check"—the operational check of the magnetos prior to flight.

magnetic   Refers to headings, courses, and such, which are measured in degrees from the magnetic north pole.

magnetic compass   A direction indicator that is sensitive to the earth's lines of magnetic force.

magnetic course   A planned track across the ground measured in degrees from magnetic north.

magnetic heading   The direction in which an airplane is pointed, measured in degrees from magnetic north.

magnetic north   The location at which the earth's lines of magnetic force appear to converge, and the reference point for the magnetic compass.

Definitions and Descriptions: A Glossary of "Pilot Talk"

magnetic variation The difference (in degrees) between true north (as shown on aeronautical charts) and magnetic north.

magneto An engine-driven "electrical pump" which supplies high-voltage electrical current to the spark plugs. Virtually all airplane engines are equipped with dual magnetos for safety and efficiency.

maneuvering speed An airspeed that provides a safe compromise between aerodynamic stall and structural overstress when operating in very rough air, or when abrupt control inputs are anticipated.

Marginal VFR A weather condition in which ceilings are expected to be between 1,000 and 3,000 feet, visibility from 3 to 5 miles. A forecast of Marginal VFR conditions should raise warning flags for all non-instrument-rated pilots.

master switch The switch used to connect the source of electrical power to the airplane's electrical system.

maximum gross weight An operating limitation which specifies the maximum weight of an airplane at takeoff.

mean sea level The average level of the surface of the ocean, and the basis for nearly all altitude references in aviation.

medical examiner A doctor who has been specially trained and then certificated by the FAA to examine aviators and issue medical certificates.

meteorology The science of weather, and the study of its effects on aviation operations.

Military Operations Area (MOA) Special-use airspace marked on all aeronautical charts to indicate the limits (both vertical and horizontal) of certain types of military activity. VFR flights may proceed through an MOA, but a great deal of caution must be exercised.

minimum controllable airspeed A training exercise in which the airplane is flown at the lowest possible airspeed without stalling. In addition to acquainting pilots with the low-speed char-

Definitions and Descriptions: A Glossary of "Pilot Talk"

acteristics of the airplane, MCA improves skills in stall recognition and recovery procedures.

mixture control A lever or push-pull device in the cockpit used to control the amount of fuel delivered to the carburetor, thereby adjusting the fuel-air ratio to the proper strength.

MOA *See* Military Operations Area.

moment The product of weight (in pounds) times arm (in inches aft of datum), used in weight-and-balance calculations. "Moment" is the *effect* of a given weight on the balance of the airplane.

MSL *See* mean sea level.

nacelle The engine enclosure of an airplane, designed for drag reduction as well as proper airflow for cooling.

nautical mile The standard distance measure for the aviation community, a nautical mile is 1 minute of latitude, or approximately 6,000 feet.

navigation The process of flying from one location to another. (*See* various types, such as pilotage, dead reckoning.)

nonstandard traffic pattern Any prescribed traffic flow which differs from the normal left-hand pattern.

NOTAM (NOtice To AirMen) Information relative to the safe and efficient operation of airplanes. NOTAMs are usually obtained through a Flight Service Station, and advise pilots of airport/runway closings, lights and radio aids out of service, and the like.

OAT (Outside Air Temperature) Usually refers to the reading of the cockpit air temperature gauge.

occluded front A weather situation in which a fast-moving cold front has overtaken and overridden a warm front; usually produces a mixture of warm- and cold-frontal weather conditions.

operating limitations Various limits (including airspeeds, weights, engine speeds) which apply to a particular airplane. The pilot is responsible for observing limitations at all times.

Definitions and Descriptions: A Glossary of "Pilot Talk"

overcast A sky condition in which 60 to 90 percent or more of the sky is covered by clouds. A Recreational Pilot may not penetrate an overcast at any time.

pattern *See* traffic pattern.

pattern altitude The height at which a particular traffic pattern is flown. In most cases, this will be 800 or 1,000 feet above the ground, but may differ in accord with local requirements or procedures.

PATWAS (Pilots' Automatic Telephone Weather Answering Service) A recorded weather briefing which saves you the trouble of going to or calling the Flight Service Station, and saves them the time required for a bunch of person-to-person briefings.

performance The description of an airplane's capabilities in terms of takeoff distance, climb rate, cruise speed, landing distance, and the like.

P factor The unique characteristic of a propeller which results in asymmetric thrust in most conditions of flight. In general, P factor tends to yaw and roll a single-engine airplane to the left.

pilotage That type of aerial navigation in which the pilot guides the airplane from one location to another by means of visual reference to prominent landmarks, with little or no preflight calculation.

pilot certificate The credential that establishes a pilot's basic ability to operate an airplane within certain limits and restrictions. A pilot certificate may be complemented by additional aircraft and/or privileges in the form of ratings.

pilot in command For general purposes, the pilot who is responsible for the safe and proper conduct of a flight; for purposes of logging flight time, it's the pilot who is actually manipulating the controls of an airplane for which he is rated.

Pilot's Operating Handbook Published by the airplane manufacturer, this document provides necessary operating information and limitations for a specific make and model.

PIREP (PIlot REPort) An advisory from an airborne aviator

Definitions and Descriptions: A Glossary of "Pilot Talk"

about weather conditions observed and/or experienced. PIREPs are an important part of the weather-information network, and should be transmitted to a Flight Service Station whenever you encounter a weather situation that might be significant for other pilots.

pitch The movement of an airplane about its lateral axis, such movement primarily controlled by the elevators.

pitot The name (pronounced "pee-toe") given to the air pressure developed as a result of movement through the air, or "ram air pressure." Pitot pressure is used to produce a reading on the airspeed indicator.

pitot-static The air-pressure sensing system which combines dynamic and static air pressure readings to operate the airspeed indicator, the altimeter, and the vertical-speed indicator.

plotter A course-measuring device used to plot the angular relationship of a planned track to magnetic or true north on an aeronautical chart.

precipitation Often shortened to "precip" by pilots, this refers to rain, snow, sleet, hail, ice pellets—any type of visible moisture which falls from the sky.

preflight inspection The routine interior and exterior check of an airplane and its equipment prior to starting the engine.

preignition The *explosion* of the fuel-air mixture inside an engine's cylinders instead of the normal rapid burning. Preignition is very hazardous to an engine's health, and must be avoided at all costs.

pressure altitude The number of feet above the standard datum plane; that is, wherever the atmospheric pressure is 29.92 inches of mercury.

prevailing visibility An official weather observer's term for the lowest visibility that covers at least half of the horizon.

primer A small hand-operated fuel pump inside the airplane that supplies raw gasoline to the carburetor to help get the engine started.

prohibited area A portion of the airspace in which no aerial ac-

Definitions and Descriptions: A Glossary of "Pilot Talk"

tivity is permitted, usually for reasons of national security. These areas are well-defined on all aeronautical charts.

propeller A rotating airfoil (lift-producing device), turned by an engine (either reciprocating or turbine) in order to produce thrust.

prop wash The turbulence created as a propeller moves large quantities of air backwards. It is relatively short-lived, and creates no major problems for following airplanes.

radiation fog Low stratus clouds formed as the result of radiational cooling, usually on a clear night. Probably the most common form of fog in most parts of the country.

range In aviation parlance, a term generally applied to distances; for example, the number of miles an airplane is able to travel with a given fuel load, or the distance of some object observed from an airplane.

reciprocal The opposite direction, in navigational terms; north is the reciprocal of south, a 90-degree heading is the reciprocal of 270 degrees.

reciprocating engine Often referred to as a "recip," such an engine is one in which linear motion of the pistons is translated into rotary motion by a crankshaft.

Recreational or Student Pilot Certificate A combination "learning permit" and medical certificate issued to a person who is working toward a Recreational or Private Pilot Certificate. If such certification is not achieved within two years of the date of issue, a new Student Pilot Certificate must be obtained.

regs Aviation slang for "regulations."

relative humidity *See* humidity.

relative wind Airflow parallel and opposite to the flight path of an airplane. The speed and direction of the relative wind plays a large part in an airplane's performance and flight characteristics.

restricted area Portions of the airspace set aside for activities which would pose a hazard to other air traffic.

Definitions and Descriptions: A Glossary of "Pilot Talk"

ridge (meteorological)   An elongated high-pressure system, in contrast to the more typical circular shape.

roll   Same as "bank," in which one wing drops, the other rises; controlled by the ailerons.

roundout   Same as "flare"—the process of altering an airplane's pitch attitude in the final stages of a landing, so as to reduce airspeed and rate of descent simultaneously.

round robin   A flight that returns to the departure point; may include stops at one or more intermediate airports.

RPM (revolutions per minute)   An expression of engine or propeller speed; in the case of fixed-pitch propellers, also an indication of the amount of thrust being produced.

runway number   The number assigned to a runway in accordance with its orientation to magnetic north. The last digit is always omitted, so that a runway lined north and south (360 and 180 degrees, respectively) would be numbered 36 for north-bound takeoffs and landings, 18 for operations in the opposite direction.

scattered   A sky condition in which 10 to 50 percent of the sky is covered by clouds. A scattered condition does not constitute a ceiling.

scud   Low layers or patches of clouds that obscure the ground, obstacles, and other airplanes. Usually applied to clouds that are distinguishable from the main body of clouds. Pilots who insist on flying low enough to stay under such clouds are known as "scud-runners," and frequently find their way into aviation accident statistics.

sectional chart   A highly detailed aeronautical chart used primarily by pilots of low-speed airplanes operating in VFR conditions.

see and avoid   The principle upon which all VFR traffic separation is based. All pilots are charged with the responsibility to maintain vigilance so as to see and avoid other aircraft.

segmented circle   A visual indicator of traffic-pattern direction at

Definitions and Descriptions: A Glossary of "Pilot Talk"

uncontrolled airports. Usually located with the windsock or wind tetrahedron, and usually lighted at night.

sequence report The hourly aviation surface weather reports collected at several hundred locations around the country.

short-field landing A maximum-performance exercise in which the airplane is brought to a stop in the shortest possible distance after clearing an assumed 50-foot obstacle at the end of the runway.

short-field takeoff A maximum performance exercise in which the airplane leaves the ground and achieves at least 50 feet of altitude in the shortest possible distance.

SIGMET Abbreviation for SIGnificant METeorological advisory; *bad news* about a weather situation such as severe thunderstorms, hail, or very high winds. A SIGMET should command the attention of *all* pilots, and usually creates a no-go situation for Recreational Pilots.

skid With reference to operations on a slippery runway or to overzealous application of brakes, self-explanatory. With reference to an uncoordinated turn in flight, a skid results when there is too little angle of bank being used—the airplane "skids" toward the outside of the turn.

slip The opposite of a skid: too much angle of bank for the rate of turn being accomplished. The airplane "slips" toward the inside of the turn. Used on occasion to lose altitude rapidly during an approach to landing.

slipstream The currents of air displaced rearward by a propeller. Same as "propwash." Also applied to the general flow of air around an airplane in flight.

soft-field landing A maximum-performance exercise in which an airplane is brought to touchdown at the lowest possible speed and sink rate. Used for landings on unprepared or rough fields.

soft-field takeoff A maximum-performance exercise in which an airplane leaves the ground at the lowest possible speed, in order to minimize the effects of traveling on a rough or soft surface.

Definitions and Descriptions: A Glossary of "Pilot Talk"

solo flight Operation of an airplane as the sole occupant, as distinguished from training operations in which an instructor is on board.

span *See* wingspan.

spatial disorientation A condition in which a pilot is unable to distinguish up from down, or whether he is turning or flying straight. Nearly always caused by the loss of outside visual clues, spatial disorientation is overcome by instrument pilots by virtue of their training to interpret the indications of the flight instruments. *Spatial disorientation in very low visibility conditions is almost always fatal for untrained pilots.*

spin In old-time aviation language, a "tailspin"—but using either name, a spin is an authoritative stall, in which the airplane heads earthward, nose down, turning rapidly. A spin is almost always the result of a stall occurring in the presence of yaw.

spinner The streamlined fairing which is often added to the propeller hub to reduce drag and improve appearance and engine cooling efficiency.

squall line An organized weather system of strong thunderstorms, usually lined up in advance of a fast-moving cold front. In general, squall line thunderstorms are the most severe of all.

stabilator An adaptation of the horizontal surfaces of an airplane's empennage, in which the horizontal stabilizer and the elevator are combined into one unit.

stability (aerodynamic) That quality of an airplane's design which results in a tendency to return promptly and easily to the condition of flight that existed prior to a disturbance.

stability (atmospheric) That quality of an air mass which results in smooth flying conditions with little or no vertical currents; if clouds form, they will most likely be of the stratus variety.

stall An aerodynamic condition in which the wing has exceeded its critical angle of attack, and has stopped producing lift. (*See* various types of stalls.)

standard day A hypothetical atmospheric situation in which, at sea level, the temperature of the air is 15 degrees Celsius and

Definitions and Descriptions: A Glossary of "Pilot Talk"

the pressure is 29.92 inches of mercury. A "standard-day atmosphere" exists at any altitude when the lapse rates of both temperature and pressure follow standard values (two degrees per thousand feet and one inch per thousand feet, respectively).

standard lapse rate *See* standard day.

standard traffic pattern The rectangular flow of traffic around an airport; used by arriving and departing airplanes as well as those remaining in the pattern for the purpose of repeated takeoffs and landings. Usually flown at 800 or 1,000 feet above the ground, and with all turns to the left.

static ports Small openings on the exterior of an airplane which admit relatively undisturbed air to the pitot-static system, to provide reference pressure for the airspeed indicator, and operating pressure for the altimeter and vertical-speed indicators.

stationary front The boundary zone between two dissimilar air masses that exhibit little or no movement across the surface. Stationary-front weather is much like that of a warm front, but somewhat less intense.

stratus clouds Layered clouds, with very little vertical development, if any at all. Generally associated with benign weather conditions, and usually productive of light precipitation, if any. Stratus clouds which form on the surface or very close thereto are known as fog.

tachometer An instrument that measures and indicates rotational speed, usually in revolutions per minute, or RPM. When used with a fixed-pitch propeller installation, the tachometer provides an indication of the power being produced by the engine.

taildragger The nickname applied to an airplane which has a tailwheel or skid instead of a nosewheel.

tailwind Air flowing from a point more than 90 degrees left or right of the airplane's heading. When flying with a tailwind, groundspeed is increased by the amount of the tailwind component.

Definitions and Descriptions: A Glossary of "Pilot Talk"

takeoff The flight operation in which an airplane leaves the surface, whether ground or water. (*See also* short-field takeoff, soft-field takeoff.)

takeoff leg *See* traffic pattern.

takeoff/departure stall A training operation in which the airplane is stalled in a configuration and condition of flight similar to that used immediately after takeoff.

taxiway The portion(s) of an airport surface intended for the movement of airplanes to and from runways.

TCA (Terminal Control Area) The airspace around a very busy airport established as "Positive Control Airspace"; definitely *off limits* for the Recreational Pilot.

three-point landing A normal landing for a tail-dragger, in which ground contact is made simultaneously on all three wheels.

threshold The edge of the paved or prepared surface which makes up a runway. (*See also* displaced threshold.)

throttle The engine control that, by means of a butterfly valve, adjusts the amount of air ingested by a reciprocating engine. In general, opening the throttle produces more power.

thrust The aerodynamic force that acts to pull the airplane forward; produced by a propeller or a jet of air, or *effectively* by the force of gravity when the airplane is moving in a descending flight path.

thunderstorm A cumulonimbus cloud or clouds that produce lightning and thunder. The high winds, turbulence, and heavy rain in and near a thunderstorm make avoidance a *must* for Recreational Pilots.

tiedown Ropes, chains, or cables used to secure an airplane when parked; also that area in an airport where airplanes are parked.

toe brakes The customary arrangement for airplane brakes, in which the rudder pedals are hinged so that toe pressure on the upper portion activates the wheel brakes.

torque A twisting force. In airplanes, torque is developed by the rotating propeller, manifests itself as a rolling tendency, and must be overcome by rudder and aileron pressure.

Definitions and Descriptions: A Glossary of "Pilot Talk"

track The actual path of an airplane across the surface; same as "course."

traffic pattern The flow of arriving, departing, and local operations at an airport. Usually consists of five segments, or "legs": takeoff (climbout on the extended centerline of the runway—also known as the upwind leg); crosswind (perpendicular to the runway); downwind (parallel to the runway and opposite the intended direction of landing); base (perpendicular to the downwind leg, and intended to position the airplane in line with the runway); and final (a descent to landing on the extended centerline of the runway).

Transition Area A cylindrical portion of controlled airspace above some uncontrolled airports, beginning 700 feet AGL and extending to 1,200 feet AGL, the purpose being to provide protection for instrument flights arriving and departing that airport.

trim Fine adjustments in control pressures so that the airplane will maintain a desired attitude.

trim tabs Small secondary control surfaces which, when adjusted by the pilot, aerodynamically position the primary control for the purpose of maintaining trim.

trough An elongated area of low atmospheric pressure; usually produces extensive layered clouds.

true airspeed Indicated airspeed corrected for variations in temperature and pressure of the air. As a rule of thumb, true airspeed increases by 2 percent of indicated airspeed for each 1,000 feet of altitude above sea level. Flight computers and calculators can provide a more accurate value.

true altitude The actual height above mean sea level.

true course/heading A course or heading plotted with respect to the location of the mapmaker's North Pole.

true north The North Pole, which lies at the convergence of the lines of longitude. Generally used for convenience in plotting courses on an aeronautical chart; "true" values must be adjusted for magnetic variation before being used in flight.

Definitions and Descriptions: A Glossary of "Pilot Talk"

turbulence (atmospheric)   Disturbances in the air, caused either by heating, rough terrain, high winds, changes in wind velocity, or all four. Turbulence results in various degrees of rough, uncomfortable flight.

uncontrolled airport   An airport with no control tower in operation, or no control tower at all.

uncontrolled airspace   Those portions of the airspace in which Air Traffic Controllers have no responsibility for separation of aircraft at any time. In general, uncontrolled airspace exists below 1,200 feet AGL away from controlled airports.

undercarriage   A British term for landing gear.

upslope fog   Fog that forms as the result of warm, moist air riding up a slope, cooling and condensing as it goes.

upwind leg   *See* traffic pattern.

usable fuel   The amount of gasoline available to the airplane's engine. Most airplanes can use all but a gallon or so of their full-tank capacity.

useful load   The weight of fuel, passengers, and baggage that may be loaded on a specific airplane. Useful load is calculated by subtracting empty weight (*see*) from maximum gross weight (*see*), and will be a different value for each individual airplane.

variation   *See* magnetic variation.

VASI (Visual Approach Slope Indicator)   A series of lights installed to the side of some runways near the approach to aid pilots in maintaining a proper, safe glide path to the runway. A VASI may also consist of single light sources of various colors, or appropriately arranged panels, or symbols painted on the runway.

vertical axis   The imaginary line which runs through the airplane from top to bottom; movement about the vertical axis is called "yaw."

vertical stabilizer   That portion of the empennage which is pri-

Definitions and Descriptions: A Glossary of "Pilot Talk"

marily responsible for an airplane flying straight through the air; acts much like a weathervane.

vertigo   In general, dizziness; with respect to pilots, vertigo refers to one of the results of spatial disorientation (*see*).

VFR (Visual Flight Rules)   Those parts of the aviation regulations covering operations when weather conditions are at or above certain minimums. Also used to describe any type of flight operation in visual conditions.

VFR Not Recommended   A phrase used by FSS weather briefers when weather conditions locally or along a proposed route of flight appear to hold hazards for non-instrument-rated pilots.

visibility   The distance one can see through the atmosphere. There are several types, two of which are of significance to Recreational Pilots: flight visibility (what you can see from the cockpit) and prevailing visibility (the lowest visibility through at least half of a ground observer's horizon).

wake turbulence   The disturbance caused by the passage of any body through the air, much like the visible wake of a boat. The production of lift by an airplane's wings lends a particularly hazardous rotational characteristic to its wake; the larger the airplane, the more dangerous its wake to smaller airplanes that follow.

warm front   The leading edge of a mass of warm air.

Warning Area   A portion of offshore airspace in which the FARs and U.S. aviation procedures may not apply. A Warning Area may also contain activity that could be hazardous to flight operations.

Weather Advisory   A warning of impending or existing hazardous weather conditions, usually broadcast in the form of an AIRMET or SIGMET.

wheelbarrowing   A most undesirable flight operation in which an airplane proceeds down a runway on its nosewheel, usually the result of too much airspeed during an attempted landing. Directional control is very difficult, at best.

Definitions and Descriptions: A Glossary of "Pilot Talk"

windmilling propeller A propeller turning with no power applied, the rotation being produced by the movement of the airplane through the air. Windmilling creates a great deal of drag.

wind shear A weather phenomenon in which the wind velocity or direction changes at various levels in the atmosphere. Can produce turbulence or derogation of airplane performance, sometimes resulting in high sink rates close to the ground.

wind sock An indicator that "streams" with the prevailing wind, showing preferred takeoff and landing direction; with experience, may be used to determine approximate wind velocity as well.

wind tee Short for "wind tetrahedron," a ground-mounted device that, due to its shape, pivots into the wind and points out the preferred takeoff and landing direction.

wingspan The distance from one wingtip to the other.

wingtip vortices The rapidly swirling portions of an airplane's wake turbulence (*see*), produced by aerodynamic forces. A vortex trails behind and below each wingtip of every airplane.

yaw The movement of an airplane about its vertical axis; the swinging of the airplane's nose from one side to the other. Generally undesirable.

Zulu A code word for Greenwich Mean Time, frequently spoken simply as "Z."

*RECREATIONAL
FLYING*

1

Flying Guidelines: The Rules and Regulations

THE FEDERAL CONNECTION

IF YOU HAPPENED to see the movie titled *The Great Waldo Pepper*, you saw, along with a lot of well-done flying scenes, a story line that illustrated the emergence of aviation regulation. As Waldo and his barnstorming associates plied their trade in the 1920s, their flying circuses introduced aviation to a significant portion of the American population; and as the activity increased, the government became increasingly concerned about the public safety. Anyone who could raise the price of a surplus World War I airplane and somehow learn (or, as in many cases, teach himself!) to fly it, was in the barnstorming business, with absolutely no regulation of his training, the maintenance of the machine, or the feats he might perform with unsuspecting citizens on board.

Once the government got into the act, of course, its regulatory powers spread into all segments of aviation, until today it is certainly one of the most governed activities in our society. That's not all bad, however, since aviation transcends state and geographic borders, and let's face it, learning to fly *safely*—with one's own kin as well as that of the folks below in mind—involves considerably more than most sporting endeavors. Someone must lay down the rules for all aviators to observe in the interest of safe use of the national airspace, and that someone is the Federal Aviation Administration.

The rules and regulations we pilots must observe have evolved into a voluminous and complicated set of publications, covering every aspect of aviation, from pilot certification to maintenance of aircraft, from definition of terms to the height of obstacles near airports, from airline operations to recreational flying—which is of course the area we're most concerned about here.

Fortunately, the relatively simple operations, the types of airplanes, and the basic objectives of recreational flying lend themselves to considerably less regulation than is required at the more advanced levels of aviation. But before we proceed further with this discussion, let's have a very clear understanding about your responsibility with regard to knowledge of the rules. There's a crystal-clear provision of the regulations that designates the pilot in command of an airplane as the person with the *ultimate responsibility and final authority as to the operation of the aircraft*. This charge has been well supported by courts across the land, and is generally interpreted to mean that *each and every pilot is responsible to know the regulations as they apply to his operations*.

For our purposes at this point, it's most important for you to realize that regulations change, for any one of a thousand reasons. We have attempted to discuss those which will probably remain essentially intact for a long time to come, but who can know the mind of the federal government? Whether it means maintaining your own set of Federal Aviation Reg-

Flying Guidelines: The Rules and Regulations

ulations (those which apply to your flying activities), or attending periodic refresher courses, or some other method of keeping up to date, *you—the pilot—are completely responsible for current knowledge of applicable rules and regulations.*

And while we're disclaiming ... we have extracted for you the fundamental portions of the flying rules, with a primary objective of helping you to understand what's expected when you're operating an airplane as a Recreational Pilot. This book is not intended to replace the regulations themselves, so you are once again responsible to obtain whatever material is necessary to make yourself aware of the complete text and scope of applicable documents.

ENOUGH OF THAT, LET'S GET ON WITH IT

The Recreational Pilot must be concerned primarily with three parts of the regulations (or "regs," as most pilots choose to call them). The first of these is Part 61 of the Federal Aviation Regulations (FARs), and has to do with the certification of pilots; the general requirements for obtaining a student permit, the areas in which you must be knowledgeable, the flight operations you must learn, and so on. Next is certain portions of Part 91, which deals solely with the operating rules—the actions and procedures expected of each pilot in the airspace, so that all can operate with confidence and safety. Finally, you must be aware of your responsibility to report an aircraft accident, and this is detailed in Part 430 of the National Transportation Safety Board's rules.

We'll deal with each of these in turn, but since many of the rules can be applied to real-world situations (and are undoubtedly better understood when placed in this light), the discussions in this chapter are limited to those parts of the regs that are not covered elsewhere in the text.

Certification of Pilots: Part 61

In the United States, pilots are issued certificates and ratings upon completion of specified tests and experience requirements. A *pilot certificate* is the basic airman credential; it provides certain privileges and imposes certain limitations. A *rating* is an official statement which, as part of a pilot certificate, sets forth special conditions, privileges, and limitations in the use of the certificate, such as multiengine, instrument, and water (floatplane) operations.

Ratings are of little concern to the recreational pilot, since the scope of training and experience provides for only single-engine airplanes with not more than 180 horsepower, and not more than four seats; the only additional rating available to the recreational pilot is that which permits him to operate that small airplane on the water.

Many recreational pilots will remain at that level of certification, renewing their credentials as required. But for those who see their aviation activity proceeding to higher levels in the certification structure, here's a quick summary of what's available:

PRIVATE PILOT This certificate removes most of the restrictions placed on the Recreational Pilot. For example, the Private Pilot may fly higher-powered airplanes with more seats, he may train further for instrument and multiengine ratings, he may use tower-equipped airports, fly at night, and fly as high as the airspace rules permit.

COMMERCIAL PILOT Required in order to accept payment for

Flying Guidelines: The Rules and Regulations

services as a pilot, the Commercial Pilot certificate comes as the result of considerable additional training and aviation experience. It is the basic credential for anyone aspiring to a career as a professional pilot.

AIRLINE TRANSPORT PILOT   In accordance with the kinds of flight operations that require this certificate (pilot-in-command of both commuter and scheduled airline aircraft), the experience, skill, and knowledge requirements are much higher than those of the other pilot credentials.

FLIGHT INSTRUCTOR   The basic pilot qualifications for becoming a Certificated Flight Instructor (CFI) include a Commercial Pilot certificate and instrument rating, plus a considerable amount of specialized training in the art of teaching.

So, the Recreational Pilot certificate may satisfy your need for aviation credentials, or it may be just the first step on the way to additional certificates and ratings. The rest of this chapter is devoted *exclusively* to the requirements and limitations of the RPC.

STUDENT PILOTS*

Eligibility, Requirements, Limitations

ELIGIBILITY

You must hold a Student Pilot certificate to begin your flight training, and there are only two things you must do to obtain it: be at least seventeen years old, and establish the fact that you are reasonably healthy. Certain doctors have

* Reference: FAR Part 61, Subpart C

been specially trained to become Aviation Medical Examiners, and have been designated by the FAA to issue the combination medical/student-pilot certificate. The minimum health standards are not very demanding, and if you meet them, you'll be issued a third-class medical certificate, valid for two years. But let's assume the worst—the doctor discovers something that might cause trouble aloft; wouldn't you rather find out about it now, before you invest in flight training, or before that medical defect causes an accident?

This medical/student-pilot certificate must be with you whenever you fly; it not only establishes your identity as a student pilot, it also provides for your flight instructor's endorsements at certain milestones in your training.

REQUIREMENTS FOR SOLO FLIGHT

One of the most trying events in the life of a flight instructor is a student's first solo flight—*unless* the CFI has done his work thoroughly and well, in which case that aerial adventure is little more than routine. (Routine for the instructor perhaps, but for the student involved, it's an event of enormous proportions!)

While the quality of work on the first solo flight will be "improvable," there's a safeguard against omissions in the quantity of training preceding that flight. The Federal Aviation Regulations contain a complete listing of the areas in which you'll be expected to demonstrate satisfactory aeronautical knowledge.

The regs also list the specific maneuvers and procedures in which you must be trained prior to solo flight. The list includes just about everything you can do with a training airplane; more important, these procedures will help to insure that you'll be able to handle nearly any situation that

Flying Guidelines: The Rules and Regulations

might come up during your early solo periods. To make sure that you've learned your lessons well, your instructor may not turn you loose by yourself until he is satisfied that you can do all these things at an acceptable level, whereupon he'll "sign you off" for solo flight.

The instructor's sign-off (or "endorsement," as it's known officially) is the visible evidence of the rather short leash on which student pilots fly in the early stages of their training. You may not operate an airplane in solo flight (when you are the only occupant) unless your student certificate has been endorsed by a CFI who has flown with you in that make and model of airplane, who has evidence that you've completed the minimum flight training requirements for solo, and who attests (with his signature) that you are competent to fly the airplane safely by yourself.

In addition to the solo endorsement on your student certificate, the instructor must make an entry in your personal pilot logbook for each training flight; if circumstances should force a temporary halt to your aerial activity, you must be reendorsed for solo flight after ninety days elapse (this is a logbook entry only).

STUDENT PILOT LIMITATIONS

The training and experience requirements for the Recreational Pilot certificate were established at an easily attainable level so that more people could share the joys of flight; but in order to help insure the continued safety of those pilots, certain restrictions and limitations had to be imposed, particularly during the training phase.

So that there will be no misunderstanding in this regard, it's worthwhile to list these limitations, just as they appear in paragraph 61.89 of the regulations:

Recreational Flying

A student pilot may not act as pilot in command of an aircraft:

1. That is carrying a passenger.
2. That is carrying property for compensation or hire.
3. For compensation or hire.
4. In furtherance of a business.
5. On an international flight (with exceptions for certain Alaskan pilots).
6. With a flight or surface visibility of less than 3 statute miles during daylight hours or 5 statute miles at night.
7. When the flight cannot be made with visual reference to the surface.
8. In any manner contrary to any limitations placed in the pilot's logbook by the instructor.

In other words, when you are flying as a student pilot, there's not much you can do except your lessons! Be patient . . . the leash gets a little longer when you leave the student ranks.

After you've jumped through all the hoops and have been certified as a full-fledged Recreational Pilot, your privileges and limitations increase somewhat. Paragraph 61.101 lays down the rules for what you may and may not do:

A Recreational Pilot may:

1. Carry not more than one passenger.
2. Share the operating expenses of the flight with the passenger.
3. Act as pilot in command of an aircraft only when the flight is within 50 nautical miles of an airport at which the pilot has received ground and flight instruction, and when the flight lands at an airport within 50 nautical miles of the departure airport.

Flying Guidelines: The Rules and Regulations

A Recreational Pilot may *not* act as pilot in command of an aircraft:

1. That is certificated for more than four occupants.
2. That has more than one powerplant.
3. That has a powerplant of more than 180 horsepower, or which has retractable landing gear.
4. That is classified as a glider, airship, or balloon.
5. That is carrying a passenger or property for compensation or hire.
6. For compensation or hire.
7. In furtherance of a business.
8. Between sunset and sunrise.
9. In airspace in which communication with air traffic control is required.
10. At an altitude of more than 10,000 feet MSL or 2,000 feet AGL, whichever is higher.
11. When the flight or surface visibility is less than 3 statute miles.
12. Without reference to the surface.
13. On a flight outside the United States.
14. To demonstrate that aircraft in flight to a prospective buyer.
15. That is used as a passenger-carrying aircraft sponsored by a charitable organization.
16. That is towing any object.

Satisfactory completion of the checkride for the Recreational Pilot certificate establishes a bench mark in your training; but until you have accumulated 400 hours of flight experience, the FAA would like to have a knowledgeable person take a look at your progress at least every year. You will have to undergo an annual flight review, which consists of at least one hour of ground instruction and one hour of flight training; the review must be accomplished and endorsed in

Recreational Flying

your logbook by an appropriately rated instructor. After 400 hours? The requirement reverts to a biennial flight review (every two years), but unless you're flying a lot, why not have your knowledge and abilities checked annually? Seems like a good thing to do.

And speaking of logbooks, the Recreational Pilot needs to keep track of all flight experience. Not only will you enjoy watching your flight time grow, your logbook is the official record of instructor endorsements, and must be carried with you on all solo flights. The 400-hour mark is important for reasons just mentioned, and the regulations require that you make at least three takeoffs and landings each ninety days to qualify as a passenger-carrying pilot.

GENERAL OPERATING RULES*

It really doesn't matter whether you are a student pilot on your very first solo flight, or an airline captain in command of a jumbo jet, Part 91 of the Federal Aviation Regulations has something for each. These general rules apply to all pilots and all aircraft in the United States, regardless of why or by whom they are being operated.

There are many sections of Part 91 which don't apply to the Recreational Pilot, and we'll omit any discussion of those parts; it makes more sense to concentrate on the rules you need to know. Also, the operating rules are involved directly with many of the flight procedures and maneuvers to be covered later in this book, so whenever a rule is more easily

---

* Reference: FAR Part 91, as applicable

Flying Guidelines: The Rules and Regulations

explained in the context of a specific flight operation, we'll discuss it fully at that time.

Right up front, Part 91 places full responsibility for the operation of an aircraft squarely on the shoulders of the pilot in command, and in the same breath, provides the pilot with all the authority needed to discharge that responsibility. If an emergency situation arises in which immediate action is required in the interest of safety, you may deviate from any of the rules to the extent required to meet the emergency. (It's possible that you could be required to justify your actions to the FAA, but that's better than getting involved in an accident or incident. When in doubt, do whatever is necessary to prevent a mishap, and fuss about the rules later.)

Nearly every set of operating rules contains some sort of catch-all, and aviation is no exception. For example, paragraph 91.5 says that the pilot in command must make himself familiar with "all available information concerning that flight." This includes weather, fuel requirements, and alternate courses of action if something goes wrong—in a few words, the regs make you responsible for being familiar with *everything* about an upcoming flight.

Catch-all number two in the aviation regulations says that "no person may operate an aircraft in a careless or reckless manner so as to endanger the life or property of another." That can cover just about any situation—when in doubt, *don't!*

Safety belts are required in all airplanes, for all occupants, for all takeoffs and landings—no exceptions. When shoulder harnesses are installed, the same regulation applies. The Recreational Pilot's trips will be of short duration, and there's not much room to stretch anyway, so why not stay buckled up the entire time you're flying?

The safe operation of an aircraft requires that its pilot be capable of rapid reactions and good judgment, both of which are impaired by alcohol and drugs. With this in mind, Part 91 prohibits pilots from flying within eight hours after the consumption of any alcoholic beverage, while under the influence of alcohol, or while using any drug which would compromise safety. The rules define "under the influence" as a blood alcohol level of more than .04 percent, and provide for stringent enforcement and severe penalties. Let common sense prevail: don't mix alcohol with your flying. (And don't carry "high" passengers; that's also against the law.)

The progress (?) of society has required yet another "substance" rule to be written into the aviation books, this one dealing with the transportation of narcotic drugs, marijuana, and depressant or stimulant drugs or substances. Airplanes have proven to be very handy and efficient in carrying these commodities, but Part 91 specifically prohibits the operation of a civil aircraft when the pilot has knowledge that drugs are on board.

It's highly unlikely that you'll ever stretch the legs of your recreational airplane far enough to be concerned about running out of fuel. But be advised that you are required by law to have enough gasoline on board at takeoff to complete the flight as planned, and still have thirty minutes' worth of fuel remaining. This would become especially important anytime you take off with less than full tanks. Don't take anyone's word for how much fuel is left—check for yourself, and be sure.

Airworthiness is at the heart of safe aircraft operations, and the Federal Aviation Administration is vitally concerned that all civil aircraft are safe for flight. As a result, an "airworthiness certificate" is issued for each individual aircraft, and it is your responsibility to see that the certificate is on board whenever the aircraft is flown. It's not enough that the

Flying Guidelines: The Rules and Regulations

piece of paper be there; it must be displayed so that it is visible to passengers and pilot. (A registration certificate, issued to the owner of the aircraft, must also be in the aircraft before flight.)

The certificate merely establishes airworthiness at one point in time, and there are a number of things that can transpire thereafter to make the aircraft unsafe to fly. Once again, the pilot bears the burden of responsibility; you are responsible for determining whether the aircraft is in condition for safe flight, and you may not operate an aircraft unless it is in an airworthy condition (that's part of the reason for thorough preflight checks, and also speaks to your knowledge of inspection requirements). In addition, the pilot is required by law to discontinue a flight when unairworthy mechanical or structural conditions occur—perhaps a rough-running engine, an oil leak, or a piece of fabric torn loose.

Every aircraft has its limitations, and the law requires that certain information be provided in the aircraft for ready reference during flight operations. Nearly all contemporary aircraft come equipped with a Pilot's Operating Handbook (there are other names for these publications, such as Owner's Manual or Flight Guide), and the limitations will be found there. But most older aircraft have sparse operating information at best, and you may have to dig a bit.

In any case, Part 91 requires that certain basic information be provided; if not in a book of some kind, then in the form of placards or instrument markings in the aircraft itself. The information may vary somewhat from one aircraft type to another, but must include operating limitations in these areas:

- Powerplant (maximum engine speed, pressures, temperatures, etc.)

Recreational Flying

- Airspeeds (maximum speed, flap extension speeds, etc.)
- Aircraft loading restrictions (weight and balance limitations)
- Flight load factors (how many Gs the aircraft is designed to withstand)

As you might have suspected, the pilot is prohibited from operating the aircraft in violation of any of these limitations. You are responsible to know what they are.

Gliders and experimental aircraft may be flown with very little in the way of instrumentation, but when you are operating a powered aircraft certificated in the standard category, there are certain minimum requirements. For VFR operations in daylight (the Recreational Pilot's environment), at least the following instruments must be installed and working:

- Airspeed indicator
- Altimeter
- Magnetic compass
- Engine tachometer, oil pressure gauge, oil temperature gauge, manifold pressure gauge (if equipped with constant-speed propeller)
- Fuel quantity gauge

Certain items of equipment are also required on board your aircraft: approved safety belts for each seat (the buckles must have a metal-to-metal latching device, and the belt material must meet certain strength standards); an approved shoulder harness for each front seat for airplanes manufactured after July 1978; an emergency locator transmitter (ELT), except for airplanes used just for training within a 50-mile radius of home base.

Flying Guidelines: The Rules and Regulations

These are the basic requirements for "ordinary" aircraft. If you are planning to operate an experimental, restricted, or homebuilt aircraft, or if you anticipate any kind of flight activities that might be construed as abnormal, obtain knowledgeable advice regarding the regulations you must observe. Interpretation of the rules can become very involved.

GENERAL FLIGHT RULES

The "crowded sky" is a myth, as you'll no doubt discover very early in your flight training. Nonetheless, pilots must observe a set of flight rules that are intended to provide a safe environment for all the users of the airspace, and that's what Subpart B of Part 91 is all about. Once again, the rules that apply directly to specific flight operations will be dealt with at the appropriate point in the text; for now, we'll look at the general flight rules which govern use of the airspace in visual conditions.

Since the underlying purpose of the flight rules is to keep aircraft from running into each other, pilots are forbidden to operate an aircraft so close to another that a collision hazard is created. Formation flying is permitted, but only by mutual consent of the pilot in command of each aircraft. A word to the wise: formation flying is a hazardous operation that requires a great deal of specialized training and skill, and should not be attempted unless and until that training is acquired.

On occasion, aircraft *do* come together in midair, and there's a better-than-even chance that one or both of the pilots involved is guilty of violating the see-and-avoid rule.

Recreational Flying

This cornerstone of the right-of-way regulations is simple enough to be quoted nearly verbatim: "When weather conditions permit, vigilance shall be maintained by each pilot so as to see and avoid other aircraft," with the obvious purpose of preventing midair collisions. In other words, when you are able to *see* another aircraft, you are expected to *avoid* it—any way you can!

The right-of-way rules are organized into logical and practical situations, with solutions to every possible confrontation. The first of these concerns an aircraft in distress, and of course everybody else is expected to get out of the unfortunate pilot's way. When two aircraft are on converging courses, the general rule states that the aircraft to the other's right has the right-of-way; and if they are approaching head-on, both are expected to alter course to the right. An overtaking aircraft must pass well clear to the right of the slower one.

There's a subset to this part of the rule, and it respects the relative maneuverability of various categories of aircraft. For example, balloon pilots can't do much about changing course, so the gas-bags have right-of-way over other types of aircraft. Gliders have right-of-way over airships (blimps), airplanes, and helicopters; and airships have right-of-way over airplanes and helicopters. Again, when you see a conflict shaping up, get out of the way and argue about who is right when you're on the ground.

When two aircraft are approaching head-on or nearly so, each pilot is expected to alter course to his right; should you find yourself overtaking a slower aircraft, give way to the right and pass well clear.

Some Recreational Pilots will become involved in aerobatic flying, and there are regulations that apply specifically to such operations. You may not perform aerobatics over any

Flying Guidelines: The Rules and Regulations

congested area of a city, town, or settlement, over an open-air assembly of persons, within a Control Zone or a Federal Airway, or at an altitude less than 1,500 feet above the surface. (An aerobatic maneuver is defined as an intentional, abrupt change in attitude, an abnormal attitude, or acceleration not necessary for normal flight.)

Just as important as keeping sufficient distance between aircraft is the matter of maintaining a safe altitude. Obviously, you'll need to operate very close to the ground during takeoff and landing, but other than that, the FAA has prescribed minimum safe altitudes for all in-flight situations except takeoff and landing.

To begin with, you must maintain enough altitude to enable an emergency landing without undue hazard to persons or property on the surface in the event of a power failure. In essence, this means that when you're flying over an area which offers no emergency landing sites except backyards and office buildings, be sure you are flying high enough to glide beyond that area if the need arises.

The rules require that when flying over any congested area of a city, town, or settlement, or over an open-air assembly of persons, you must maintain an altitude of at least 1,000 feet above the highest obstacle within a horizontal radius of 2,000 feet.

Over other than congested areas, you may fly as low as 500 feet above the surface if you like; over open water or sparsely populated areas, you must keep your aircraft at least 500 feet away from any person, vessel, vehicle, or structure. Flying at 500 feet may be legal, but it's not necessarily smart. There are few flight operations which justify a Recreational Pilot getting that close to the ground.

Finally, since Recreational Pilots are rather severely restricted by weather conditions, the basic VFR weather minimums are made very easy to remember: whenever the visibility on the ground is less than three miles, *don't fly!* If you should find yourself flying into an area of lowering visibility, turn around and go back to where you can see safely.

The Federal Aviation Regulations cover in great detail just about every conceivable operation that a pilot can conduct. But when it comes down to what really counts, your memorization of the regs is not as important as your resolve to use a lot of common sense in all of your flight activity. When something doesn't look right, doesn't feel right, it probably isn't. It's usually wise to back off, take another look, and see if there isn't a better way to get the job done.

Before we leave this section on general flight rules for Recreational Pilots, remember three things: (1) rules (especially those written for an activity as dynamic as aviation) are subject to change; (2) pilots are responsible for knowledge of the regulations (changes and all!) that apply to their flight operations; and (3) you must consult a complete set of the regulations to understand what we've treated very briefly in this chapter. You should check with your Flight Instructor about the various ways to stay up-to-date with the Federal Aviation Regulations.

Airplane Structures and Components

WE'RE GOING right back to square one in this chapter, to the very fundamentals of airplane construction and design. The objective is not to make you an expert, but to provide at least a basic understanding of why your flying machine is built the way it is. This should help you to carry on intelligent conversations with fellow pilots and mechanics, as well as explain your new hobby to nonpilot friends.

BASIC STRUCTURES

At the very minimum, an airplane needs a wing to provide lift, control surfaces to manage that lift, an engine, landing gear, and a place for the people to ride. There are always

Recreational Flying

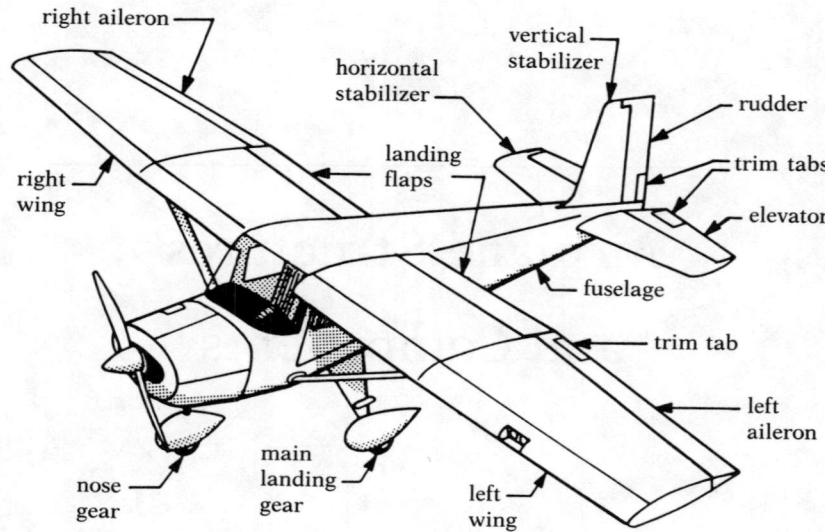

Airplane components and structures

strength and weight considerations, since every pound of structure means that more power will be required to make the machine fly.

This was a very real problem for the early airplane builders, since the technology of the times limited would-be aviators to wooden construction. The Wright brothers could do no better than a wire-braced wood truss structure (as a matter of fact, their work was heavily influenced by an engineer who specialized in railroad bridges!), which was remarkably light and strong, but the web of wires created a great deal of drag. As the technology evolved, airplane structures began to be covered with cloth, then with sheets of lightweight metal.

Most of today's light airplanes are fabricated entirely of aluminum—curved inner structures to which sheet metal is

Airplane Structures and Components

riveted or bonded to form smooth, streamlined shapes. There is a rapidly growing population of molded plastic and fiberglass airplanes, which may well be the wave of the future in airplane construction.

It is important for pilots to realize that virtually every part of an airplane is subjected to some of the loads encountered in flight. Therefore, even a dent in the aluminum skin needs to be checked by a competent mechanic or inspector to make sure that it won't affect the flight characteristics of the machine. This is one of the reasons for a thorough preflight inspection.

BASIC AIRPLANE COMPONENTS

The Fuselage

The main body of an airplane is nearly always in the shape of a long, tapering cone, something like a spindle; early French aviators used the word "fuselage" to describe what they saw, and the name stuck. The fuselage serves to streamline, it's the attachment point for other airplane structures (like engines, landing gear, wings, and tail surfaces), and it provides creature comforts—most noticeably to protect you from wind and weather!

Fuselage construction techniques haven't changed much over the years, and the airplanes you'll encounter as a Recreational Pilot will probably be either truss (flat or nearly-flat surfaces covered with cloth) or semimonocoque (sheet metal riveted or bonded to flat or curved surfaces). The former method imposes nearly all of the load on the truss itself, while the semimonocoque design transfers a great deal of the stresses of flight to the skin itself. (There are still a few molded plywood airplanes around, and some of them may employ a full-monocoque construction, in which the entire load is carried by the skin.)

Bonded surfaces, fiberglass, and plastic materials are becoming very popular because they are very light and very strong. The lack of rivets and joints on the surface of an airplane also contributes considerably to its efficiency in flight, and there will be more of these "new-method" airplanes in our skies as time goes on.

Wings

These are the primary lift-producers, by virtue of the curved shape which accelerates the flow of air over the top side. Most airplane designers have incorporated fuel storage in the wings, and in some cases a place for baggage as well. Most retractable-gear airplanes pull their wheels into recesses in the wings. But the big job is to serve as the primary airfoil, developing lift.

Designers' imaginations have produced unnumbered wing designs over the years, but we can categorize the most popular types rather easily:

NUMBER OF WING SURFACES

MONOPLANE/BIPLANE/MULTIPLANE Most airplanes today have but one wing (monoplane), extending equally left and right from the fuselage. In order to derive more lift from a given wing length, some airplanes are built with two wings (biplane), one stacked above the other. And with the philosophy that "if two is good, more is better," there have been two-, three- and many-winged (multiplane) flying machines. (It should be noted that the amount of drag increases rather remarkably with the addition of more wing surfaces, and most of these designs were doomed to failure on that count.)

WING POSITION

HIGH/MID/LOW/PARASOL   As you view an airplane from front or rear, the wings will be mounted either on top of the fuselage (high-wing), in the middle of the fuselage (mid-wing), or on the bottom of the fuselage (low-wing). A parasol wing is mounted above the fuselage by means of struts—the airplane literally hangs from the wing. While there are some personal preferences among pilots as to which style is "better," there is precious little difference in performance or technique required to fly them.

TYPE OF WING BRACING

EXTERNAL STRUTS   In this type of construction, the wing is braced against both upward and downward loads by means of wooden or metal struts, and there are usually wires present as well. Makes for a relatively lightweight structure, but the penalty is paid in appearance and drag.

CANTILEVER BRACING   By far the most popular method in use today, the cantilevered wing has no external braces, but depends on the strength of an inner spar (one or more beams running from wingtip to wingtip) to carry the load. The benefit is streamlining and beauty, the cost is weight.

Control Surfaces and Flight Controls

A rudder, an elevator, and a pair of ailerons make up the "primary control surfaces" on most light airplanes. They are primary because their operation by the pilot directly affects the movement of the airplane about its three axes; they are the primary means by which you will manage the forces at work on the airplane, and therefore control its flight.

Recreational Flying

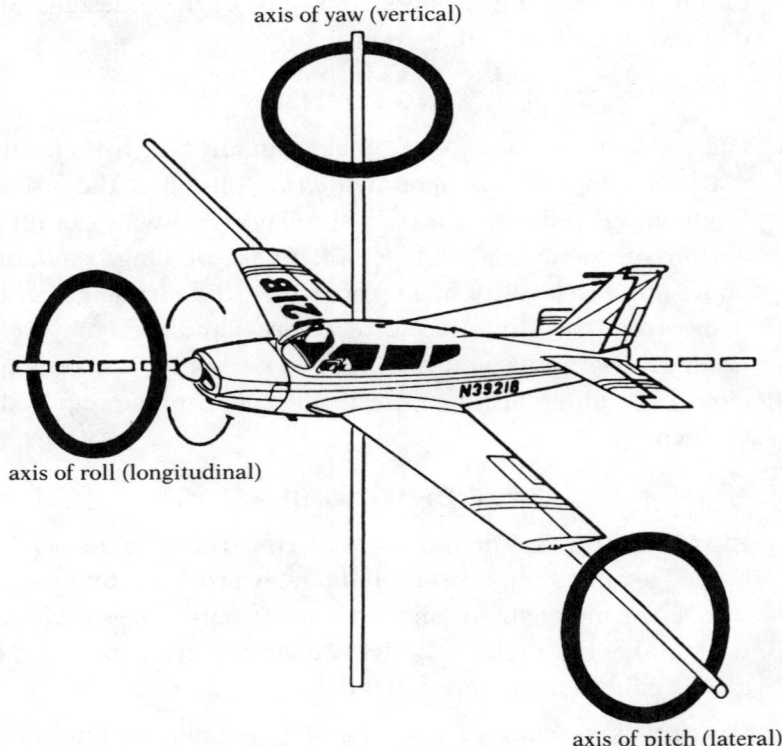

Three axes of rotation

The rudder, linked by cables or push-rods to a pair of pedals in the cockpit, moves from right to left in response to pedal pressure, and causes the nose of the airplane to move right-to-left accordingly. It is usually hinged to, and is essentially a part of, the fixed vertical stabilizer.

The elevator (it appears to be two units, one each side of the rudder, but they are interconnected and move as one) is responsible for up and down movements of the airplane's nose. When the pilot applies back pressure to the control stick or

Airplane Structures and Components

wheel, the elevator moves upward and causes the nose to do likewise; the opposite is true when forward pressure is exerted. Not unlike the rudder, the elevator is also hinged to a fixed surface—in this case, it's the horizontal stabilizer.

Rudder, elevator, and the fixed stabilizers are nearly always attached to the fuselage at the extreme rear to gain maximum aerodynamic leverage from control movements. These two control surfaces (and the stabilizers) are known collectively as the "empennage," another aviation word borrowed from the French.

At the trailing edge of each wing, near the tips, is a pair of control surfaces known as ailerons, or "little wings." They move in opposite directions when the control wheel is turned (or when the contol stick is moved from side to side), and provide roll control. Ailerons are rigged so that when one moves up, the other moves down, and the resultant imbalance of forces causes the airplane to tilt, or bank, from side to side as the pilot desires.

A secondary control surface—a trim tab—is sometimes installed for the purpose of relieving control pressure when the primary surface must be held in a position other than streamlined with the airflow. A trim tab is operated by a crank or switch or lever in the cockpit; it is hinged to the trailing edge of its primary surface, and moves in the opposite direction. You'll find elevator trim tabs on nearly all airplanes, with tabs on rudder and ailerons added as airplane size, complexity, and power increase.

Engine and Propeller

Light-airplane engines have evolved over the years into a very standardized product. For all practical purposes, there are only two engine manufacturers, and the differences between their engines are minuscule.

In most small airplanes used by Recreational Pilots, engines are horizontal-opposed and air-cooled (that is, the cylinders lie on their sides on either side of the crankcase, and there's no radiator as in your car), operate on the four-cycle principle, and normally drive the propeller directly, no gears involved. Airplane engines are usually housed in a nacelle, which improves the looks, the streamlining, and the flow of cooling air around the engine.

There are some survivors of other engine types still flying, most notably the radial powerplants, so called because the cylinders are arranged in a starlike pattern around the crankshaft. If you become interested in ultralights or certain homebuilt airplanes, it's likely you'll become acquainted with two-cycle engines, and who knows, we may one day see small turbine (jet) engines in light airplanes.

Propellers are the thrust producers in an airplane's propulsion system. Little more than sophisticated fans, propellers are long, thin airfoils which produce lift when moved through the air; in this case, the "lift" is produced in a horizontal direction, and the resultant force pulls your airplane forward. Props may be made of metal or wood, may have two, three, or more blades (although anything more than two blades is unlikely on a small airplane), and in high-power installations, may be capable of changing pitch (blade angle) while rotating for more efficient thrust production.

In any case, propellers must be considered the most hazardous appliance on the airplane, at least in the sense of ground operations. With safety in mind, think of a propeller as a rotary guillotine, and stay clear! Whenever there's an encounter between a person and a propeller, the prop wins every time.

Airplane Structures and Components

Landing Gear

An airplane requires some way to get around on the ground, and some means of moving relatively fast down the runway for takeoff, to say nothing of the need for a smooth transition from sky to earth at the end of a flight. Enter the landing gear, which answers those needs in various configurations.

Predominant in our aeronautical population is wheeled landing gear, but if you go north to Canada or Alaska you may be surprised at the number of airplanes operating on floats or skis. There are even combinations of these ("amphibians," just like frogs), for pilots who need the flexibility of operating the same airplane from two or more different surfaces.

In the beginning, airplanes sat level upon the ground, mostly because the propellers were mounted in back. As more powerful engines came along, propeller blades had to be made longer to absorb the additional horsepower, and the now-classic tailwheel, or "taildragger," configuration showed up. Today, the tricycle-gear (little wheel up front) airplane is considered to be conventional, with tailwheel types relegated to a rather small segment of the population.

Some airplanes are equipped with retractable landing gear. There are benefits in speed (much less drag when the wheels are tucked away inside fuselage or wings) and appearance, but of course there's a price to be paid: the initial cost of a more complicated system, higher maintenance bills, and weight.

Airplanes don't have springs in the sense that automobiles do, but bumps in the runway and the shock of landings are smoothed in various ways. There are air-oil shock absorbers that operate on a hydraulic-piston principle, spring-steel landing gear legs, and systems that soak up shock by compressing rubber "doughnuts" or by stretching elastic cords.

Stopping power—wheel brakes—is usually applied by foot pressure on the hinged upper part of the rudder pedals. In some airplanes this results in automotive-type brake shoes being pressed against the revolving wheel, in others the pedals generate hydraulic pressure which then does the work. Hand levers are still to be found in the brake systems of some older airplanes, and require their own brand of technique to be used smoothly and properly. Most contemporary systems incorporate disc brakes, which are much lighter and more reliable. And in some cases—the very old airplanes—there's no braking system at all, except a lot of back pressure on the stick to press the tail skid down and literally bring things to a grinding halt!

3

Basic Aerodynamics

THE STUDY of the various forces exerted on an airplane by its movement through the air can get to be very complicated—witness the graduate-level courses in aerospace engineering at nearly all universities, and the unbelievably complex computer programs that control today's jet aircraft.

But the study of aerodynamics ("aero" for air and "dynamic" to indicate movement) can be greatly simplified for the Recreational Pilot. Your primary consideration is to understand that there are four forces—lift, thrust, drag, and weight—acting on every airplane in flight, and that you will need to manage three of them. (There's not much to be done about weight once you're in the air!)

A pilot (Recreational or otherwise) can't do his job properly, efficiently, smoothly, or safely unless he understands at least the basic aerodynamics involved.

THE FOUR FORCES

Consider, at the outset, an airplane in flight, and being operated so that there is no change in direction, no gain or loss of altitude, and no change in speed—a condition generally referred to as "straight and level, unaccelerated flight."

In order to achieve this state of affairs, the pilot has used his flight and engine controls to manage the four forces of flight.

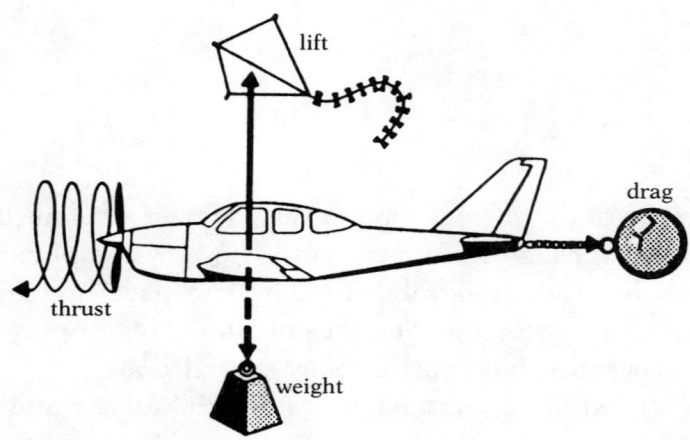

Forces acting on an airplane in flight

The forces are represented by arrows: *thrust* pulling the airplane ahead, *drag* holding it back, *lift* pulling it upward, and *weight* pulling it down. Notice that the four forces act in pairs, thrust versus drag, and lift in opposition to weight. Each force needs to be considered by itself:

THRUST   Developed by the propeller as it rotates, thrust pulls the airplane forward, providing the speed that enables the wings to do their work. Since the propellers in most light airplanes are driven directly by the engine, an increase in engine speed will increase propeller speed, and therefore increase

Basic Aerodynamics

thrust. As you'll see later, this may be translated into an increase in either altitude or speed.

DRAG   This force is created by the very form and shape of the airplane as well as certain reactions of the airstream flowing around it. In a very simple airplane (no wing flaps, fixed landing gear), drag is relatively constant and doesn't require much in the way of management by the pilot. However, when the amount of drag can be changed significantly (by the extension and retraction of landing gear, flaps, and so on), the pilot has yet another tool to work with to make the airplane do his bidding. In any event, drag always acts to retard the forward progress of the airplane.

LIFT   The very essence of flight, lift makes it all possible by overcoming the force of gravity (weight) and sustaining the machine in the air. Lift is produced as the result of forward motion, and therefore depends heavily on the airplane's speed, which depends heavily on the force of thrust, which depends heavily on the force of drag . . . and you can begin to see how the four forces are completely interrelated.

WEIGHT   Everything on the earth is attracted toward the center of the sphere, and when that attraction is measured, we call it "weight." Airplanes are no exception, and each flying machine weighs a certain number of pounds; that weight must be overcome by lift in order to achieve flight, but since the airplane is subject to some movements not encountered by earthbound vehicles, its weight can "change." This phenomenon and its effect on flight will be discussed in more detail very shortly.

LIFT PRODUCTION

All four of the aerodynamic forces are important to the flight of an airplane, but lift needs to be treated in more detail.

Nearly all of the maneuvers and flight operations you'll perform involve the management of lift with one or more of the flight controls.

When an airplane's wing is at rest—no forward movement—and there's no wind blowing to move air over it, the *pressure* of the air is the same on top and bottom surfaces. Equal pressure all around means that there is no force trying to move the wing in any direction.

But add movement, and as soon as air begins to flow around the wing, a unique physical change takes place. The curvature of the wing's upper surface causes the airflow to speed up somewhat, and when this happens, the pressure of that accelerated air is slightly reduced. The higher-pressure air below the wing will now exert an upward force, tending to push (lift) the wing toward the area of lower pressure. It's a lot like a tug-of-war, when one side weakens a little bit: if the forces are out of balance, something's gotta give.

An airplane's wing also deflects a considerable amount of air downward, and the resultant reaction contributes an additional upward force—more lift.

The faster the wing is moved through the air, the more the pressure is reduced on top and the more the air is deflected downward. Imagine your airplane rolling down the runway, traveling faster and faster, lift increasing all the time. At some point, the force of lift will equal and then exceed the weight of the airplane, and *flight* happens!

This is a very basic principle, and always works; in other words (all other conditions being the same), whenever you cause the airplane to move faster through the air, the airflow around the curved wing surface and the deflection will increase and produce more lift.

As you might have guessed from the "all other condi-

Basic Aerodynamics

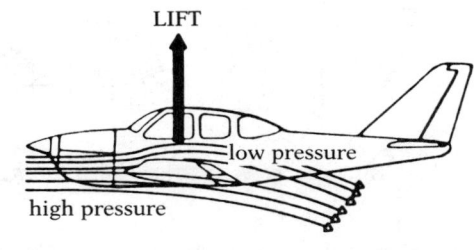

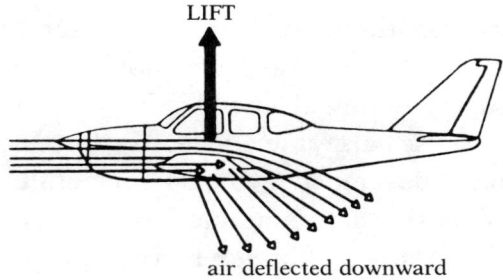

Lowered pressure and downward deflection generate lift

tions" in that last statement, there are other factors that affect the production of lift. There are different wing shapes, and devices that alter the characteristics of the wing, and methods of changing the airflow patterns; but when considering the small airplanes most likely to be used by the Recreational Pilot, the most important "other" condition is the angle at which the wing meets the oncoming air—the "angle of attack."

In the next illustration, the first wing is shown with a small angle between its midline and the direction of the air through which it is moving; the second wing has been inclined slightly upwards, the third angled up even more.

The angle is usually increased by your operation of the elevator controls, but of greater importance is your understanding that as the angle of attack increases, lift increases.

Recreational Flying

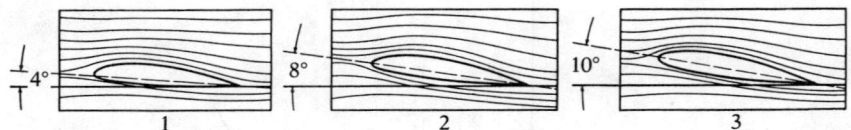

Lift increases with angle of attack

When the wing is rotated upward, the curvature of the upper surface is effectively increased, meaning that the air must move farther, and therefore faster, in order to travel the slightly increased distance; and we know that when the airflow speeds up, pressure drops. At the same time, the increased angle causes a greater deflective force.

Now we have developed two basic principles relating to the production of this aerodynamic force; lift is increased as both *airspeed* and *angle of attack* increase. (The opposite is also true—decreasing angle of attack and airspeed reduce lift.) There are a thousand different combinations of these which will produce various amounts of lift, and we'll get to that later in this discussion.

That's the good news. . . . The bad news is that continuing to increase the angle of attack for more lift reaches a point of diminishing returns. If you can imagine a wing turned upward so far that it is *perpendicular* to the airflow, you can see that there's little or no difference between the distance the air must travel no matter which way it goes. There is no difference in pressure, therefore no lift is produced. The wing is "stalled."

In the real world, this condition—an aerodynamic stall—occurs at a very low angle of attack, something on the order of 15 degrees or so for most light planes. As the angle is increased, a point is reached where the difference in distance is so great that the airflow can no longer move in smooth

Basic Aerodynamics

layers over the top of the wing; it "separates," and the flow becomes turbulent.

This is known as "burbling," and is the same kind of behavior you see in the rapids of a fast-moving stream flowing over an uneven bottom.

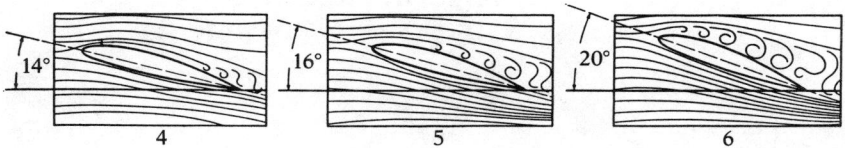

Higher angles of attack result in burbling, and the wing stalls

From the pilot's point of view, there's one important thing about an aerodynamic stall: when it happens—for whatever reason—lift production stops, or at least falls off remarkably. Now the forces are out of balance again, weight takes charge, and some altitude will be lost.

THE FOUR FORCES RELATED

An airplane in flight is affected by lift, weight, drag, and thrust; the relative strength of these four forces has a great influence on the airplane's behavior. For example, when thrust and drag are equal, the airplane will neither speed up nor slow down—it is flying at a steady airspeed. But when thrust is increased (such as when you increase engine speed) to a value greater than drag, the airplane will accelerate. If you reduce thrust (throttle back), the airplane will slow down because drag is now the greater force. In similar fash-

Recreational Flying

ion, when lift is increased the airplane will gain altitude, and will descend whenever lift is reduced to a value less than the weight of the airplane.

FLIGHT CONTROLS: MANAGING THE FORCES OF FLIGHT

Suppose, just for a moment, that your airplane is a "flying wing": no fuselage, no empennage, just a wing in straight, level, unaccelerated flight. If this wing were speeded up, there would probably be a pitch change (nose up or down) due to the movement of the center of pressure, an aerodynamic phenomenon which is beyond the scope of this discussion. As a matter of fact, the pitch change would likely become uncontrollable, so a small horizontal tail—a "stabilizer"—was added, mounted as far behind the wing as possible to provide leverage.

Airplanes should nose over a bit and glide in the event of engine failure, so designers have purposely made them nose-heavy. To counteract this tendency, and allow the airplane to fly level in normal circumstances, the horizontal tail is actually an upside-down wing; when air moves around it, a downward force is created. Now, with the wing-tail combination moving through the air at a given speed, the aerodynamic situation looks like this (see figure opposite).

The arrow pointing up represents the lift being produced by the main wing, the downward-pointing arrow represents the built-in nose-heaviness, and the arrow under the tail represents the "down-lift" created by the upside-down wing. When the two down arrows are equal, the system will main-

Basic Aerodynamics

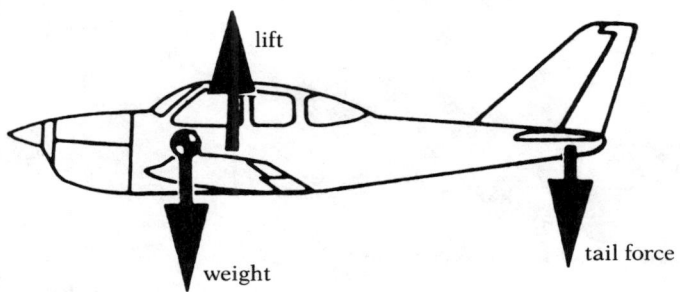

The stabilizer offsets an airplane's built-in nose-down tendency

tain level flight—the tail surface has indeed become a horizontal stabilizer.

Unfortunately, such a system creates as much of a problem as it solves. The forces are in equibrium only when the wing-tail combination is moving through the air at a certain speed. When airspeed is increased, lift is increased on both main wing and stabilizer; the airplane will begin to gain altitude, and there will also be a nose-up pitch change as the stabilizer exerts its influence.

The net result of all this aerodynamic activity is a climb whenever thrust is increased, and descent whenever thrust is reduced (everything works in reverse order as thrust changes). The wing/horizontal-stabilizer system operates at only one airspeed, and altitude is controlled by the amount of thrust the pilot applies.

To fly at a *different* airspeed, it is necessary to control the downward force created by the horizontal stabilizer. This is done by adding a moveable surface to the trailing edge of the stabilizer, and connecting it with cables or rods to the control wheel or stick so the pilot can move the surface up or down.

When this surface is moved upwards (with back pressure

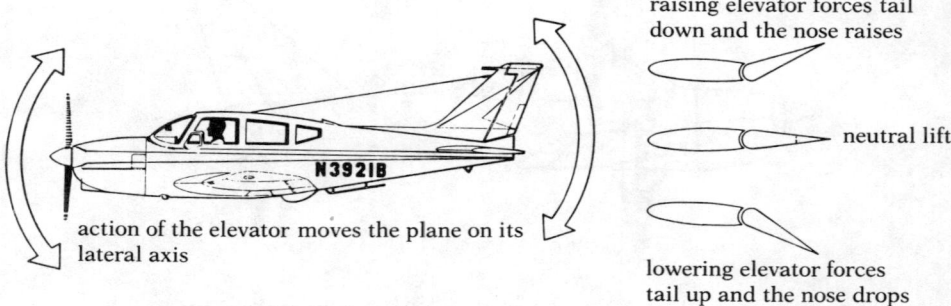

Effect of the elevator on an airplane's pitch movements

on the wheel or stick), the air flowing under the stabilizer must travel farther to get around the increased curvature, which lowers the air pressure on the underside, pushing down on the tail, and raising (elevating) the nose of the airplane. The reverse is true when the elevator is moved upwards.

Now, when you want to fly more slowly, first reduce thrust a bit, simultaneously applying enough back pressure on the control wheel or stick to raise the nose. How much? Just enough to increase the angle of attack so that lift production will be maintained at a value equal to weight, and the airplane will be flying level at a new, slower airspeed. If your objective is to fly *faster*, you must add thrust and lower the nose enough to reduce the lift created by the higher speed.

While the principle is fresh in your mind, let's consider flying level at progressively lower airspeeds. Thrust is reduced in small amounts, pitch attitude being increased as required to produce lift equal to airplane weight. At some point in the sequence, the angle of attack will reach a point where airflow over the top of the wing cannot continue in a smooth, layered fashion, and a stall will occur.

Basic Aerodynamics

There *is* a lower limit to the speed at which any airplane can fly!

All airplanes incorporate a similar stabilizer/control-surface arrangement for vertical movement, the swinging of the nose from left to right (this movement is called "yaw"). The vertical stabilizer has the same shape on both sides so it won't exert any aerodynamic force in straight flight, and acts just like a weathervane whenever air pressure is greater on one side or the other. The *rudder* is the moveable portion, and is connected to the rudder pedals so that the pilot can create or prevent yaw as needed.

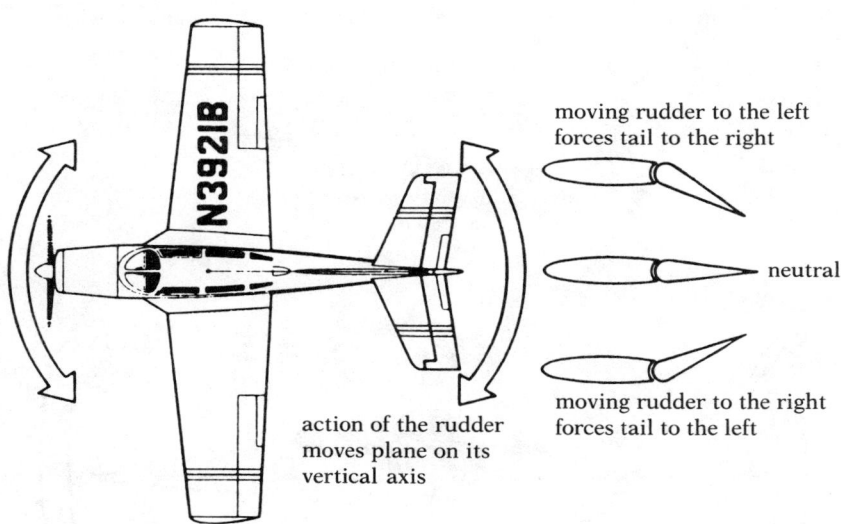

Effect of the rudder on an airplane's yaw movements

The third major control surface is the pair of ailerons, which serve to alter the lift produced by the wing itself. The ailerons are rigged to move in opposite directions, so that when the control wheel or stick has been turned or pushed to the right, the left aileron moves down and the right aileron up, and vice versa.

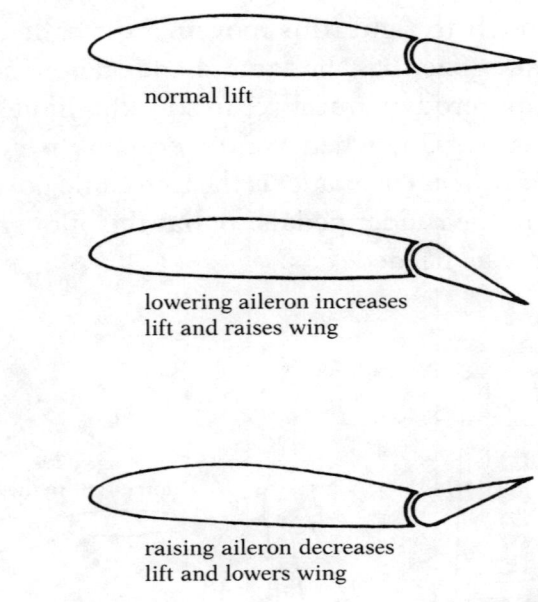

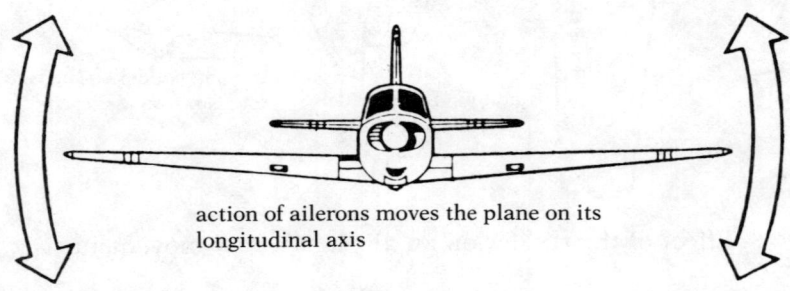

Effect of the ailerons on an airplane's roll movements

The curvature of the upper wing surface has been increased in the area affected by the left aileron, which means

Basic Aerodynamics

that lift over that segment of the wing has been increased somewhat. The opposite is true of the right wing, where the up aileron reduces lift a small amount. When the controls are operated in this manner (left aileron down, right aileron up), the left side of the wing produces more lift than the right, and a roll to the right results. This is also known as a "bank."

By including the engine controls (with which you adjust power, and thereby change thrust) with elevator, rudder, and ailerons, the pilot has the tools to manage all the directions of movement of which an airplane is capable: climb or descend, change airspeed, roll, yaw, and pitch. These flight-management tools will of course be used in various combinations to achieve or maintain the desired attitudes and conditions of flight; but keep firmly in mind that in any case, you are dealing with small changes in air pressure, and your inputs should be smooth, small changes in control pressures.

Trim tabs are secondary control surfaces that act to change the curvature (and therefore the lift production) of the respective primary control. For example, when you are flying at a low airspeed, it's necessary to hold the nose up; to relieve the constant wheel or stick pressure, the pilot of an airplane equipped with an elevator trim tab (nearly all airplanes have one) operates the trim control to move the tab in the opposite direction of the primary control deflection.

The slight increase in airflow over the top of the elevator creates a small amount of lift, and when the tab is moved to the point at which you feel the wheel or stick pressure disappear, the elevator is properly trimmed. The trim tab will hold the elevator in the desired position until there's a change in the aerodynamic situation. A good pilot always flies his airplane in trim.

Recreational Flying

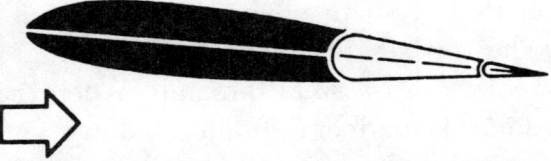

elevator in the neutral position

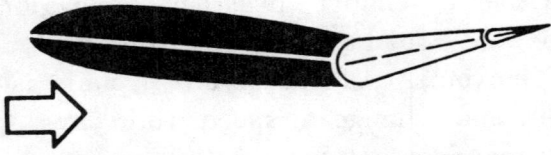

up position of the elevator is required to hold the nose in the level flight attitude

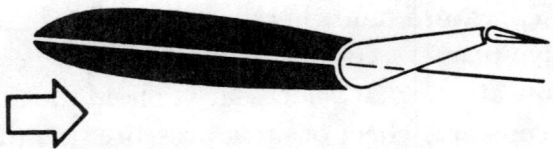

trim tab must be adjusted downward to hold elevator in this position to relieve the pressure on the control wheel

Trim tabs are used to relieve pressure on the control wheel or stick

Basic Aerodynamics

PROPELLER EFFECTS—THE "FIFTH FORCE"

Until such time as jet engines replace propellers on light airplanes (don't hold your breath until *that* happens!), pilots will have to live with the aerodynamic effects of those whirling blades. This discussion truly belongs right here in the section on basic aerodynamics, because a propeller is in fact a rotating wing, shaped with one curved surface, and it develops lift when it rotates through the air. The major difference, of course, is that a propeller's lift is expressed in a horizontal direction, and is more accurately referred to as "thrust."

In one of the rare conveniences of aerodynamics, the major propeller effects of concern to the Recreational Pilot act in the same direction; when they're at work, you'll notice a tendency for the airplane to yaw and roll to the left. One effect at a time, then.

Virtually all American-made airplane engines turn clockwise (as viewed from your seat, behind the engine), which means that a clockwise spiralling motion is imparted to the strong flow of air moved rearward by the propeller. Follow the spiralling slipstream along the fuselage and you'll notice that it always strikes the vertical stabilizer on its left side, tending to push the tail to the right, nose to the left.

The slipstream effect is greatest at low airspeeds and high power settings (such as during takeoff, climb, go-around, and stall recovery), and is always countered by liberal application of right-rudder pressure. How much? Whatever is needed to prevent yaw.

The second significant effect is generated by a combination of propeller rotation and changes in the angle of attack. With the airplane in level flight and the plane of propeller

Recreational Flying

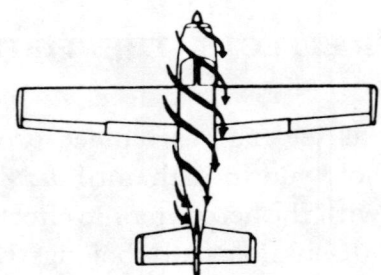

The spiralling slipstream pushes the nose to the left

rotation perpendicular to the propeller shaft, each blade will produce the same amount of thrust.

Now the problems begin. Your airplane doesn't always fly with the prop shaft parallel to the line of flight—as when, for example, you are flying slowly, with the nose held higher than normal.

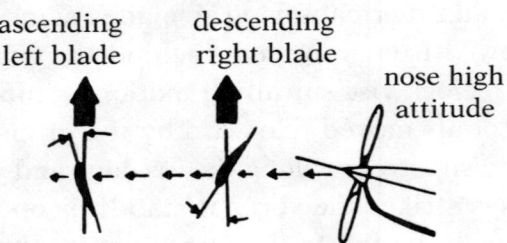

P factor pulls the nose to the left

Because the blades are fixed firmly to the shaft, there must be a change in the angle of attack when this entire assembly (airplane, engine, prop shaft) is tilted upwards. The illustration shows how much the angle of attack of the right-hand (descending) blade has increased. On the other

Basic Aerodynamics

side, the angle of the ascending blade has gotten smaller. The result is unequal, or asymmetric, thrust. The right-hand side of the propeller disc is pulling harder than the left, causing the airplane to yaw left. You should expect asymmetric thrust in nose-high, high-power situations, and the proper corrective action once again is rudder pressure—as much as required to prevent the yaw.

Gyroscopic effect and torque effect (twisting) are present in any piece of machinery that incorporates a rotating mass, but the weight of a light airplane propeller is not enough to introduce significant problems in these areas. Nevertheless, you should consider all of these characteristics which are unique to single-engine, propeller-driven airplanes, anticipate the effects (in general, left yaw and left roll in high-power, low-airspeed, nose-high situations), and apply whatever control pressures are necessary to overcome them. You've done your job properly when the airplane doesn't roll or yaw if you don't want it to!

4

Basic Flight and Engine Instruments

IT'S VERY CLEAR that airplanes can be flown without fancy and expensive instrumentation; after all, the earliest aviators had virtually nothing in that regard. (If you look closely, however, you'll see a short piece of string attached to one of the struts of the Wright brothers' first airplane; they were able to judge their airspeed roughly by observing the angle of the string in the wind.) For many years after rudimentary instruments became available, pilots continued to gain a great deal of information from the wind in their faces and the feel of flight. In other words, they were flying "by the seat of their pants."

Today, the Federal Aviation Administration requires certain minimum instrumentation in all airplanes, mostly for reasons of safety. Since the airplanes you'll fly as a Recreational Pilot are included, this chapter deals with the

Basic Flight and Engine Instruments

principles of operation and basic uses of the airspeed indicator, the altimeter, the tachometer, and the manifold pressure gauge; we'll spend a lot more time on the magnetic compass in the navigation chapter, and detail some additional engine instruments in the discussion of airplane powerplants.

THE AIRSPEED INDICATOR

The speedometer on your car is a rather simple mechanical device; it measures the rate at which the wheels are turning, and presents the information in terms of the number of miles you would travel in an hour at that rate. An airplane's wheels obviously stop turning very shortly after lift-off, so some other method of determining speed is required.

We mentioned the Wright brothers' use of a string; although primitive, it was effective, but more important, it permitted them to visualize the effect of air pressure as the airplane moved along. The faster they flew, the straighter the string. You can get a graphic demonstration of the pressure principle by putting your hand out the window of a moving car; the faster you go, the more pressure you feel trying to push your hand backward. Since there is a direct and reliable relationship between speed through the air and pressure exerted, the accepted method of measuring a light airplane's velocity is to measure the air pressure created by forward movement.

Somewhere on one of the extremities of your airplane (most likely near the end of one wing, to avoid the propeller slipstream) is the pitot tube, no more than a small metal tube with its open end facing directly into the line of flight. As the airplane moves through the air, the pitot tube samples the pressure thus created, and directs it to the airspeed indicator on the instrument panel.

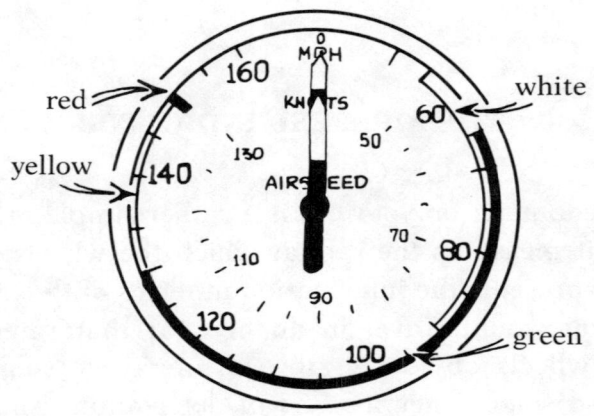

A typical airspeed indicator, with colored bands
to show airspeed ranges and limitations

The heart of the airspeed indicator is a bellows that expands as the pressure increases, and contracts when the pressure drops. Made of thin, flexible metal, this bellows is connected to the needle on the face of the indicator; as air pressure increases and expands the bellows, the needle is moved to indicate higher numbers—higher speed. As you slow the airplane, the bellows moves towards its deflated position, indicating lower speeds.

Even on the ground, with absolutely no airplane movement, the airspeed indicator will show a positive reading when the pitot tube is facing into the wind. Fortunately, the

Basic Flight and Engine Instruments

airplane's wings are also sensing the wind, and are already developing some lift—a happy situation, since it permits flight with a shorter takeoff roll.

All of this simplicity carries a price. You need to understand that air becomes less dense as you climb higher in the atmosphere, and that lower density means lower pressure. (If you can grasp the idea that moving your hand through water is a lot more difficult than moving your hand through air, you should have a handle on the difference in pressure between dense air and less dense air.)

An airplane that is moving through the air at a true speed of 60 miles per hour will show its pilot an airspeed indication of 60 under standard conditions; that is, air of a certain density, exerting a certain pressure in the pitot tube. But fly that same airplane at the same speed at a level in the atmosphere where the air density is considerably lower (therefore exerting less pressure), and the airspeed indicator will show a false reading; the bellows can't be inflated as much by lower-pressure air, even though the airplane is traveling at the same speed as before.

This problem is solved to some extent by providing a vent in the airspeed instrument case so that the bellows is always working against the lowered pressure of the outside air in flight. Nevertheless, you should count on a difference between true airspeed and indicated airspeed in almost every flight situation, with true airspeed usually the higher number.

The airspeed indicator will be affected by gusts of wind while you're flying, and since the instrument has a rather small-scale dial, you will find it difficult to take accurate readings on occasion. This is a minor problem, because the airplane's *attitude* is more important, and when the attitude is correct, the proper airspeed will result.

THE ALTIMETER

There are two ways of saying it—some prefer the accent on the first syllable, AL-ta-meet-er, while the more generally accepted pronunciation is al-TIM-a-ter—but by any name, the altimeter does nothing more than measure air pressure and provide an indication of the altitude that corresponds to that pressure.

An altimeter contains a bellows very similar to the one in the airspeed indicator, the major difference being that the altimeter's bellows is sealed. It is therefore responsive to every change in atmospheric pressure. The bellows is housed in a case connected to the outside air by a tube, the open end of which is located somewhere on the airplane in a region of undisturbed air, usually on the side of the fuselage toward the rear, and usually in the form of a small, round metal plate with several drilled holes.

For the sake of discussion, let's assume that when the bellows is sealed, it contains sea-level pressure. The manufacturer arranges the mechanical connections between the bellows and the pointers on the face of the instrument so that it reads 0—sea level. Now, whenever the air pressure in the case changes, the bellows will contract or expand, causing the pointers to indicate a change in altitude; higher numbers for lower pressure, and vice versa.

If you took delivery of a new altimeter from the sea-level factory, and flew home to your airport located 1,000 feet above sea level, the instrument should indicate 1,000 feet when you land. There are other factors that may affect the accuracy of an altimeter (changes in pressure due to changing weather systems or abnormal temperatures, for

Basic Flight and Engine Instruments

example), and you will find a small adjustment knob with which you can reset the altimeter once on the ground at home.

The changes will be rather small (except when there's a *massive* change in the weather), but you must get in the habit of adjusting the altimeter reading to agree with the airport elevation (above sea level) prior to every flight. Aviators who operate in the lower levels of the atmosphere (below 18,000 feet) speak a common altitude language—everyone uses sea level as the base for altitude measurements.

Technology is gradually replacing the old-style three-pointer altimeters with digital readouts, but it's a safe bet that all the airplanes in the Recreational Pilot community will have altimeters with three hands. Fortunately, only two of these are of concern. The indicating system is composed of pointers on a circular dial, which is marked in 20-foot increments.

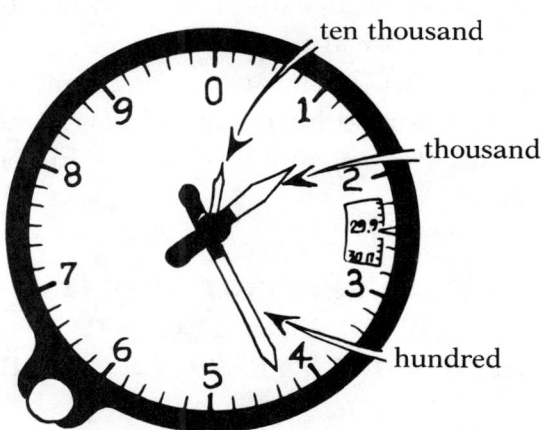

A typical three-pointer altimeter, indicating an altitude of 1,435 feet

As you begin a climb from sea level, the longest pointer will begin to move clockwise to show an increase in altitude. Look carefully, and you'll see the next-longest pointer also moving, but at a rate one-tenth that of the longer one; by the time the longest pointer has reached 1,000 feet (one complete circle on the dial), the shorter pointer rests on the number 1, indicating 1,000 feet. This 10-to-1 ratio continues throughout the range of the instrument, so that at any given time, you can determine your altitude in thousands of feet by reference to the middle-sized pointer, and can refine that reading to within 20 feet by noting the position of the longest hand on the altimeter dial.

The shortest pointer also begins moving when you climb, but at a rate which is one-tenth that of the thousand-foot pointer. When you've climbed high enough to cause ten full revolutions of the longest hand (a climb of 10,000 feet), the shortest pointer will rest on the number 1.

The adjustment knob on most altimeters moves not only the pointers, it also changes the numbers displayed in a small window, usually on the right side of the altimeter dial. When a pilot has knowledge of the local "altimeter setting" (aviation talk for local atmospheric pressure in inches of mercury adjusted for field elevation), that number should be inserted in the pressure-adjustment window. With a proper altimeter setting, the instrument should read very close to airport elevation.

BASIC ENGINE INSTRUMENTS: THE TACHOMETER AND THE MANIFOLD PRESSURE GAUGE

Unlike automobiles, airplanes don't have the "idiot lights" that let a potentially serious situation develop before

Basic Flight and Engine Instruments

they let you know that something's wrong in the engine room. Because even the smallest airplane engines operate in a demanding, rapidly changing environment, and the consequences of failing to notice an impending problem can be disastrous, the FAA has seen fit to mandate certain minimum engine instruments, ones which provide a "right now" picture of engine condition.

Our objective at this point is to explain only those engine indicators that speak to the amount of thrust being used, hence the limitation of this discussion to the tachometer and the manifold pressure gauge.

Most of the airplanes used by Recreational Pilots are equipped with fixed-pitch propellers, made of one solid piece of metal or wood. These props are normally driven directly—that is, there are no gears between engine and propeller—and the rotational speed of both components is the same. With this installation, engine speed is measured in revolutions per minute, commonly shortened to RPM, and is displayed to the pilot on a tachometer, a round dial with a single pointer. Open the throttle, thereby causing the engine to turn faster, and the tachometer shows an increase in RPM. Most tachometers read in hundreds of RPM, with 100-RPM increments. In general, an increase in RPM, as displayed on the tachometer, indicates an increase in thrust.

Some Recreational Pilots may operate airplanes equipped with constant-speed propellers, in which a fairly complex system twists the blades as they rotate, in order to maintain whatever engine speed the pilot selects. With the obvious elimination of the tachometer as a thrust indicator (RPM remains constant even though the engine is delivering more or less power), such installations require an additional instrument to show the pilot where he stands, thrust-wise.

The manifold pressure gauge does the job; it is also a

round dial, on which a pointer moves across a scale of pressure readings. The standard scale shows inches of mercury, in one-inch increments, and as the throttle is opened (admitting more air to the engine through the intake manifold, hence the name), the engine develops more power. Although propeller RPM will not change (within the limits of the system), additional thrust has been generated, and the manifold pressure gauge is the primary indicator. It is important to note that there are very clear-cut limits in the relationship of manifold pressure and RPM; the airplane's operating limitations must be consulted to prevent engine damage.

5

Basic Flight Operations

AT THE OUTSET, it may appear that making your airplane go through all the gyrations necessary for normal flight is an impossible task. Your first flight or two as a student may indeed cause you to wonder if you can chew gum and walk at the same time, to say nothing of coordinating the movements of both feet and both hands in order to control a machine that seems determined to move in all three directions at once!

Take heart, because an airplane is *expected* to move up or down, left or right, and forward at the same time during certain manuevers; but no matter how complicated those maneuvers become—even if you move on to aerobatics—they are all made up of four basic flight operations. You can fly straight and level, climb, descend, or turn—and develop a thousand combinations to make the airplane do whatever is necessary for the situation at hand.

Our objective in this chapter is to analyze each of the basic flight operations, and provide some tips and techniques for learning to accomplish these building blocks of flight. In

Chapter 9 and subsequent chapters, you'll learn how to combine the basics into useful applications of your flying skills.

STRAIGHT AND LEVEL FLIGHT

Once past the drills and exercises of your student experience, you'll spend more time in straight and level flight than anything else, so it pays to understand it well. Surprisingly, you may find that flying straight and level properly and consistently is one of the most demanding of all the basic flight operations. That's because altitude, heading, and airspeed should remain constant when you're doing it right, and it's always easier to make things change than to keep them the same. All of this can be handled if you understand what's going on, and what you are expected to do.

For training operations (as well as most cross-country work), most pilots choose a power setting of approximately 65 percent, which will produce a reasonable airspeed in level flight. That's one of the constants to be used for this operation, joined by a selected altitude and heading.

After level-off from either a climb or descent (the leveling technique will be discussed shortly), after the throttle has been adjusted to produce the proper RPM, *pitch attitude* becomes the key to maintaining a constant altitude. It's going to be cut-and-try for awhile until you discover just how much space to put between the horizon and the cowling of your airplane to fly level, but the importance of using that visual cue cannot be overemphasized. Don't try to "fly the altimeter"—using the cowling as your reference, make small changes in pitch attitude until the altimeter stands still. Once you've discovered where the nose should be, you can always return to that attitude, and the altimeter will do your bidding.

When the level-flight attitude is established, use the elevator trim tab (move it s-l-o-w-l-y!) to relieve any forward or back pressure required to hold the nose exactly where you want it. Should gusts of wind or vertical currents in the air cause a small gain or loss of altitude, apply wheel or stick pressure as required to make the small pitch-change attitude changes that will keep the altimeter steady; on a turbulent day, you're going to be busy! (As this slight up-and-down process is taking place, engine speed will vary somewhat due to the changes in air load imposed on the propeller. Unless the changes are gross, live with the variations, chalking them up to the inconvenience of bumpy air.)

Most light airplanes are very stable (that is, resistant to change) in yaw, and you'll find that very little rudder pressure is required to keep the nose from moving to the left or right in cruising flight. But remember that strong propeller force trying to yaw the nose left during climb? Most manufacturers provide some help by mounting the vertical stabilizer with a slight offset. When the airplane reaches cruise speed, invariably higher than that used for a climb, the offset becomes more effective, resulting in a tendency to yaw to the right in cruise. If you have to hold a bit of left rudder pressure to keep everything in order, so be it. At any rate, your job is to prevent yaw—don't let the nose swing.

Roll (bank) control involves keeping the wings level with respect to the horizon, and may require some constant pressure on the wheel or stick. You'll notice this more when flying solo in airplanes with side-by-side seating, because there's more weight on the left side. The strongest tendency is to steer the airplane as if it were an automobile, using the control wheel or stick to prevent yaw. In this situation, the airplane will wind up flying in a slip, which is uncoordinated, uncomfortable, and inefficient—it sets up a lot of unneeded drag.

Rather than steer, use visual cues once again to do the job right; prevent yaw with rudder pressure, and keep the wings level (relative to the horizon) with aileron pressure.

As you become better acquainted with your airplane, you will discover the idiosyncrasies that exist in all machines; with airplanes, it's often a matter of accepting a little bit of bank, a slight amount of rudder pressure, to maintain straight and level flight. If the pressures required are more than a little, you should probably have a mechanic take a good look at the rigging of control cables and surfaces. (But of course, you'll never know there's anything wrong unless you know how things *should* look and feel!)

CLIMBING FLIGHT

You can rest assured that the airplanes used by Recreational Pilots will require all the power their engines can produce in order to get off the ground in a reasonable distance. And in virtually every case, full power is also recommended for climbing; these little airplanes just don't have many horses to spare, and there's no damage to be done by running the engines at full power for relatively long periods of time. (Check the limitations, of course; some engines may have time limits on full-power settings.)

Given a constant amount of thrust for climbing, you must decide on an airspeed to use, and there are several factors to consider. First, you need to see over the nose in your search for other airplanes, so you don't want a low airspeed that

Basic Flight Operations

requires a high pitch attitude; but if you lower the nose sufficiently to provide good forward vision, the angle of attack may be so small that very little climbing takes place. Second, the engine is cooled by the flow of air around it, and as airspeed decreases, engine cooling may suffer.

A compromise is in order, and you'll find it in the airplane's handbook, under "best rate of climb airspeed" (some of the older airplanes don't have handbooks, and you may have to search elsewhere for this number). Also known as V_y, this airspeed has been found to produce the greatest possible rate of climb—the maximum gain in altitude per unit of time. V_y is a function of angle of attack, and your job is to discover the pitch attitude—indicated by a reliable visual cue, such as a cowling-horizon reference—that produces that airspeed. Once discovered, this attitude should always be used for a normal climb, unless you notice problems of forward visibility or engine cooling, in which case you should change the attitude as required for the situation at hand.

A companion climbing speed is V_x, or "best angle of climb airspeed." At this speed (also in the airplane handbook), the airplane will gain the maximum altitude per unit of distance; V_x is important when you need to clear an obstacle very close to the end of the runway, or anytime you need a rapid short-term increase in altitude. Once again, you should determine the pitch attitude for the best angle of climb airspeed, and store it away in your memory bank of visual cues. Operating at best angle of climb airspeed is an abnormal procedure used most frequently for short-field takeoffs, and we'll talk more about that in Chapter 19.

In either case—best rate or best angle—there's a pitch attitude that will produce the proper airspeed, and when you wish to enter a climb, it's simply a matter of readjusting the throttle to provide full power, then smoothly placing the nose where it belongs on the horizon. Don't try to "fly" the airspeed indicator, or you'll find yourself oscillating all over the sky; rather establish, maintain, and adjust the pitch attitude using your reliable visual reference, and airspeed will take care of itself. *Airspeed is merely the indicator of the proper pitch attitude.*

Yaw control during climb is largely a matter of realizing that P force is at its greatest in this condition of flight, and you will need to apply a footful of right rudder pressure in order to keep the nose from moving to the left. Same old technique does the job: keep the wings level with aileron pressure, prevent yaw with rudder pressure, and the airplane will do its best work for you.

Leveling off from a climb can be accomplished with power or pitch changes, or a combination of the two. Bear in mind that you are climbing by virtue of additional thrust; that is, thrust which is not required to maintain altitude at climb airspeed. It follows, then, that if you wish to stop the climb, simply reduce power to the point where there's just enough to maintain level flight. Such a procedure works, but you will level off at the climb airspeed; economical, to be sure, but you won't cover many miles that way.

More likely, you'll want to cruise (fly level) at a significantly higher airspeed, so the following procedure is recommended; just before the desired cruising altitude is reached (perhaps 50 feet or so for most light airplanes), slowly press the nose down toward the horizon, and let the excess thrust translate into additional forward speed. As the airplane moves faster, several things take place: the engine will speed up because of the reduced load, lift will increase because of the additional airflow over the wings, and the airplane will probably try to yaw to the right because of the reduction of P force. There will be a moment or two during this transition

Basic Flight Operations

from climb to cruise when you will be very busy keeping all of the forces and effects under control; practice and recognition of the visual cues are the keys to success here.

DESCENDING FLIGHT

Just as entering and sustaining a climb requires a change in thrust, a descent takes place when the lift-weight relationship is unbalanced in favor of gravity. In other words, when power is reduced to a value less than that required for level flight, the airplane will, if left alone, begin to descend at a rate directly related to the amount of thrust taken away, and at the same airspeed for which it was trimmed in cruise. So, one way to enter a descent is to simply reduce power, let the airplane assume its own trimmed pitch attitude, and watch as the altimeter unwinds.

There is no published "best" airspeed for a descent, but there are at least three considerations: engine cooling, the effect on your (and your passengers') ears, and the structural speed limits of the airplane. First, a sudden and significant reduction of power followed by a higher-than-cruise-airspeed descent means less internal heat, and more external cooling, which may be more than the metals in the engine can cope with. Second, when humans descend at rapid rates, the change in air pressure often manifests itself as discomfort, if not outright pain, in the ears—some folks just can't swallow fast enough.

Third, every airplane has speed limitations that must be observed if you intend to keep the covering on the wings, and the wings on the airplane! In the Operating Limitations section of the handbook, you'll find both a "never-exceed" speed

(V_{ne}) and a "caution range" of airspeeds, the latter to be observed in turbulent air.

Your Flight Instructor will help you determine comfortable, efficient speeds for descents under different conditions, but in any event, there are several aerodynamic effects you should be aware of. While a simple power reduction permits the airplane to descend at the trimmed airspeed, and therefore requires little if any pitch inputs from the pilot, there will be a change in P force, and you'll notice a definite need for some left rudder pressure to keep the nose from swinging to the right.

If you choose to descend at an airspeed other than that for which the airplane was trimmed in cruise, you will need to provide the appropriate pitch input—either forward or back pressure on the control wheel or stick, as required. Establish the new pitch attitude, and adjust the trim to hold it. The pitch change (and the accompanying airspeed change) may or may not cancel out the change in P force; do whatever is necessary with the controls to make the airplane do what you want.

During a descent, an airplane with a fixed-pitch propeller will tend to increase engine speed as the air density increases, so you will need to adjust the throttle setting from time to time in order to maintain a constant RPM.

Levelling off from a descent usually involves nothing more than replacing the thrust that was taken away. This readjustment of the throttle should begin about 50 feet above the desired level-off altitude, and if the objective is level flight at your normal cruise airspeed, simultaneously increase power and adjust the pitch attitude to the values you established previously. If you'd rather level off at a higher- or lower-than-normal airspeed, prepare for another transition period during which pitch, power, and yaw inputs must be adjusted to find the right value for each.

TURNING FLIGHT

Up to this point, much has been made of keeping the wings level, and preventing yaw. These are major objectives of straight flight, whether level or in a climb or descent. But you've got to change direction sooner or later, and in an airplane, there's more involved than simply turning a wheel or pushing a tiller, as you would in a car or boat. Remember that your airplane travels *in* its medium, not on it. There's nothing such as a highway or the surface of a lake for an airplane to push against in order to turn.

To be sure, if you were to apply rudder pressure and cause the nose to swing across the horizon until it's pointing where you want to go, the airplane would eventually fly in that direction. But since it moves *through* the air, it would skid sideways for a considerable distance before resuming normal flight on the new heading. The transition would be uncomfortable (you'd feel a great deal of sideways force) and not very efficient (whenever an airplane is forced to fly un-streamlined, drag builds up rapidly and significantly). There must be a better way.

In order to understand the whys and wherefores of the proper technique for turns, consider a rear view of a typical airplane, with total lift represented by a large arrow acting vertically through the center of the fuselage (for now, lift is equal to weight, making for level flight).

Recreational Flying

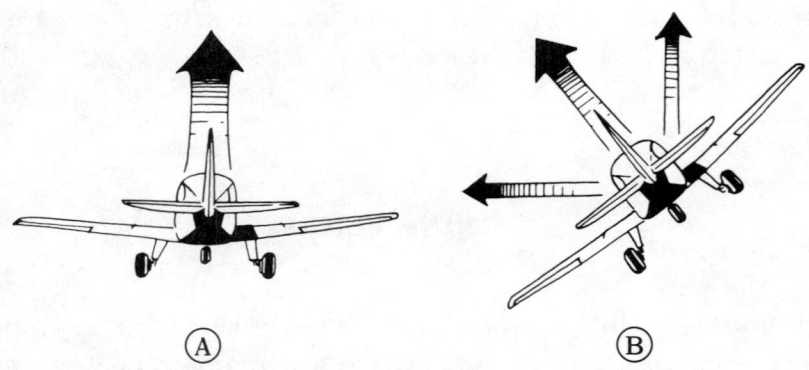

Vertical lift in straight flight (A), and its redistribution during a level turn (B)

When you apply pressure to the control wheel or stick so as to displace the ailerons, lift production increases on one wing, decreases on the other, resulting in a rolling motion. For straight flight, we've concentrated on *preventing* roll, but now, the name of the game is "bank the wings so that the airplane will turn."

With even the slightest hint of a bank, there's an important change in the distribution of lift. Instead of a single lift vector acting straight up, a small portion of that vertical lift now acts in the direction of the bank, causing the airplane to be pulled sideways. The size of the pull depends on the angle of bank.

Try to visualize the airplane in the illustration above as if it were not moving forward through the air (an impossibility, since no motion means no lift, but try!). Can you see that the sideways force would actually pull the airplane to the left? Now resume the forward motion in your mind's eye, and the result will be a curved path through the sky, with the airplane still streamlined, still moving through the air in a

Basic Flight Operations

very efficient manner, and with no uncomfortable side loads on its passengers. There *is* a better way!

In essence, then, turning an airplane properly involves use of the ailerons to redistribute the lift of the wing, so that some of the lifting force takes on the job of pulling the airplane sideways. You will likely be amazed to discover the extremely small amount of bank required to produce a noticeable turn; and that's why it is so important to keep the wings level when you're trying to fly straight.

You might have suspected that this is a "good news, bad news" situation, and you're absolutely right. Since the total lift didn't change when the airplane was rolled into a bank, the vertical component of lift will be reduced by the amount of sideways force developed.

In the absence of other inputs, an airplane will always lose altitude when turning. But it doesn't have to be that way, because you have complete control over total lift production, and if you don't want to descend while turning, simply increase the wing's angle of attack with some back-pressure on the control wheel or stick. How much? Whatever it takes to make the altimeter stand still.

Unfortunately, another culprit shows up during a turn, this one known as centrifugal force. It's present whenever any body is moved in a circular path, and it is always trying to move that body away from the center of the circle. You've seen (and felt) centrifugal force at work in an amusement park ride, where the floor of a revolving drum falls away and the passengers are held tightly against the side; remember how heavy you felt?

The very same force is at work on your airplane (and everything in it) whenever you roll into a turn, and the amount of centrifugal force increases more rapidly than the angle of bank. Although normal training maneuvers don't involve much centrifugal force (you'll barely notice it), the effect on the airplane must be considered. In a bank of 60 degrees (in which the airplane is rolled until the wings are at an angle of 60 degrees to the horizon), centrifugal force has effectively *doubled* the weight of the airplane. The wings must now produce twice the amount of lift (in addition to the side component used for the turn), which must be generated by additional airspeed or an increase in angle of attack, or both.

Centrifugal force is labeled a "culprit" because *the increase in effective weight always causes the airplane's stall speed to increase.*

Yaw control is every bit as important when turning as it is in straight and level flight. In yet another "no free lunch" situation, yaw is induced by the very act of positioning the ailerons for a turn. The aileron that moves upward is displaced into the low-pressure area on top of the wing, while its companion on the other side moves downward into an area of relatively high pressure; and where the pressure is higher, the air is more dense and produces more drag. Therefore, the aileron that moves down creates a dragging force which holds that wing back and causes yaw.

This phenomenon is known as adverse yaw, and it must be overcome by—you guessed it—whatever rudder pressure is required to keep the nose from swinging opposite the intended direction of turn. Most contemporary airplanes are designed so that adverse yaw is practically nonexistent, but you can bet your boots that the older ones will yaw away from turns with a vengeance! A spin is lurking just around the corner if a stall occurs in this condition of flight.

Proper turning flight requires a lot of learning, both in control technique and in developing your own set of visual

Basic Flight Operations

cues so that you'll know when you are doing it right. When you begin, use the same angle of bank every time, and begin to assemble a mental-image library of just how things look. As you roll into or out of a turn, use rudder pressure to prevent the nose from swinging in the opposite direction, and develop a good set of reference points to help you with pitch control throughout the turn.

Airplane Engines: Principles and Operation

THE AIRPLANE'S POWERPLANT gets it off the ground, makes climbing possible, and, quite literally, sustains flight. Simply because the powerplant is the most essential component of your airplane, it behooves you to get familiar with how it works and how you can keep it working longer. Airplane engines and propellers are also very expensive—you do yourself a financial favor by treating them right.

You should consider the engine and propeller as a combination of components that provide propulsion for flight. The engine develops power to turn the propeller, which, because of its aerodynamic shape, creates thrust. For all practical purposes, you can consider power and thrust synonymous in most light airplanes.

Given the importance of long-term, dependable operation of the powerplant, we can go no farther without making you aware of the various limitations and restrictions applied to the operation of various systems. You must refer to, and abide by, the instructions set out by the manufacturers of both engine and propeller as installed on the airplane you are flying.

PRACTICAL FEATURES

Virtually all light airplanes are powered by reciprocating, horizontally opposed, air-cooled engines. It's not necessary for the Recreational Pilot to become intimately familiar with those descriptors, but a basic understanding will help you do a better job of operating any airplane's powerplant.

A reciprocating engine is one in which linear (straight-line) motion is converted to rotary motion; in the typical light airplane installation, you'll find pistons moving back and forth within cylinders, and by virtue of their connection to a common crankshaft, that linear motion is put to work turning the propeller.

A bicycle rider is a reciprocating engine; fuel (leg muscles) drives two pistons (feet) in a basic up-and-down motion which turns the sprocket and propels the bike. For streamlining and improved cooling, light-airplane engines are arranged with the pistons (usually four of them) lying on their sides, to the left and right of the crankshaft—thus, "horizontally opposed." (If you can imagine the aforementioned cyclist lying on his back and pedaling, you're getting close to the concept.)

The requirement for airplane powerplants to be very light in weight (in relation to power output) accounts for these engines being air-cooled. You'll notice that the cylinders are finned to provide more surface area to dispel the heat of combustion, and your preflight inspection includes a check for obstructions in the airflow pattern inside the engine cowling. There's not much difference between normal operating temperature and destructive heat in an airplane engine.

For the same reason—minimum weight—there are no diesel engines in light airplanes (or any airplanes, for that matter). Rather, the typical powerplant is a four-cycle engine fueled by aviation gasoline (avgas). A mixture of fuel vapor and air is drawn into a cylinder (intake stroke), whereupon a piston compresses the mixture (compression stroke); it is subsequently ignited by an electrical spark, and the rapid burning creates pressure within the cylinder, forcing the piston to move and thereby turn the crankshaft (power stroke). In the final portion of the four-cycle engine's operation (exhaust stroke), the residue of combustion is forced out of the cylinder by the return action of the piston.

The amount of fuel-air mixture admitted to the engine controls its speed, and therefore its power output; when the engine is driving a propeller, more power means more thrust.

Four strokes of the piston complete the operating cycle of a typical airplane engine

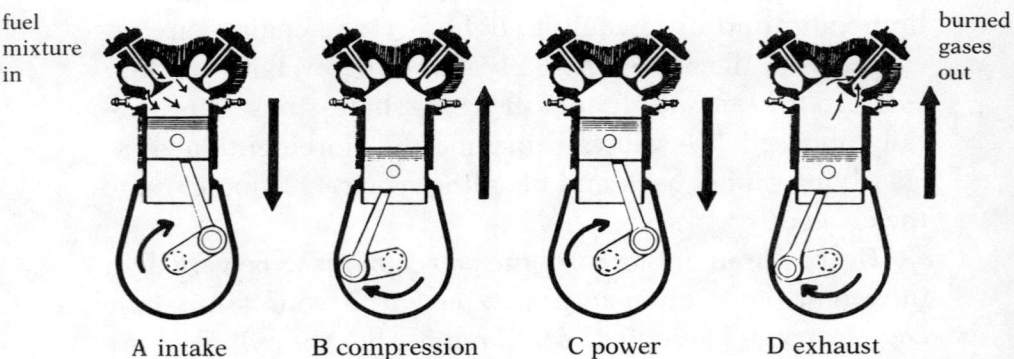

Airplane Engines: Principles and Operation

The throttle control in the cockpit lets you adjust a butterfly valve in the intake line for this purpose.

Of equal concern for the operator of an airplane engine is the quality of the mixture, since only a relatively narrow range of fuel-air mixtures will support efficient combustion. Although the air is less dense aloft, there is little change in the amount of fuel being fed to the engine, resulting in an overrich mixture. If no adjustment is made, engine power will suffer because the mixture is too rich to burn properly. A mixture control in the cockpit provides a way to reduce the amount of fuel selectively, thereby maintaining the proper fuel-to-air ratio, and optimizing engine operation.

OIL AND FUEL SYSTEMS

The airplanes used by Recreational Pilots are powered by relatively simple engines; that is, they are mostly low-powered, have four cylinders, and are normally aspirated (not fuel-injected). Despite this simplicity, there must be a system for lubrication, a means of storing fuel and delivering it to the engine, and certain engine instruments required by law.

A small oil tank (known as a sump) is usually incorporated into an airplane engine, from which oil is supplied to lubricate all the moving pieces. Some of the lubrication is accomplished by merely allowing oil to splash onto the parts, and some is done by forcing oil under pressure through very small lines throughout the engine's innards. Oil is the lifeblood of your airplane engine; oil pressure and temperature gauges are required by the regulations, and provide in-cockpit indications of that aspect of engine health.

One of the most important tasks during the preflight inspection is a check of the oil quantity; since it is responsible for considerable internal engine cooling as well as lubrication, you must be certain that the proper amount of oil is present before starting the engine.

Small-airplane engine simplicity is also manifested in the fuel system, which is frequently nothing more than a tank in each wing, an ON-OFF valve in the cockpit, and the lines necessary to let gasoline flow from tank to engine. (Many airplanes, especially the newer ones, have a fuel selector which permits you to use fuel from left or right tanks as desired.) A high-wing airplane has a built-in gravity-flow system, but those with low-mounted wings usually depend on electrical pumps to create adequate fuel-supply pressure.

In any event, the Federal Aviation Regulations require that your airplane be equipped with some sort of indicator to make you aware at all times of the amount of fuel in the tanks. Most fuel gauges are little different from those found in automobiles, and some of the old-timers have nothing more than a cork float in the tank with a wire sticking up through the cap to let you know how much gasoline is left . . . that's simplicity!

Gasoline must be vaporized and mixed with the proper amount of air in order to burn inside the engine and provide power. The device that accomplishes this rather important job is the carburetor, which is located in the air intake line, and operates on the principle of reduced air pressure. In light airplanes being flown at low altitudes, carburetion is largely an automatic process, with little pilot attention required.

Two cockpit controls—the throttle and the mixture

Airplane Engines: Principles and Operation

control—provide for management of the fuel supply to the engine. The throttle is a simple butterfly valve in the intake line; opening the valve (by pushing on the throttle) admits more of the fuel-air mixture to the engine, and it runs faster. The mixture control, a lever or plunger in the cockpit, operates a valve in the fuel supply line and regulates the amount of fuel flowing to the carburetor.

The engine operates best on a fuel-air mixture of approximately 14 parts air to 1 part fuel, and the carburetor will supply this mixture strength at all power settings from idle to full power with no adjustment required. But when an airplane climbs, say, to 3,000 feet MSL or higher, the change in air density requires a change in mixture strength; the pilot must reduce the amount of fuel to maintain the desired 14-to-1 ratio, an operation known as leaning the mixture.

When you have established the airplane in normal cruising flight, slowly move the mixture control rearward, keeping in mind that at "full rich," there's too much gasoline, and that if you move the contol all the way and close the fuel valve, the engine will quit. Somewhere in between is the ideal mixture strength for this altitude. As the fuel-air ratio is changed, engine speed will increase slightly (you'll see this on the tachometer, and hear it as well), and when the peak RPM has been reached, the mixture strength has been adjusted properly.

If you happen to be learning to fly at a high-altitude airport (more than a couple of thousand feet or so above sea level), you may be required to lean the mixture before takeoff, to accommodate the less-dense air in which you're operating all the time. Your Flight Instructor will teach you to fine-tune this leaning procedure so that you get the most out of the engine in every situation.

ELECTRICAL SYSTEMS

Since Recreational Pilot operations are limited to daytime flying in VFR conditions, there is no need for lights, flight instruments, radios, or other systems which rely on electrical power. However, most training airplanes are equipped with electrical starters, and some will have wing flaps and other electrically powered units, so a discussion of basic electrical systems is in order.

A storage battery rotates the engine for starting, and provides an emergency source of electricity; once the engine is running, a generator or alternator picks up the load. These units should be thought of as "electrical pumps" which energize the electrical system whenever they are turning, and which replenish the battery as required.

All airplane engines, even those with full-fledged electrical systems, are equipped with a totally separate system to provide the spark that ignites the fuel-air mixture in the engine. In order to keep this vital source of energy in operation at all times, a pair of magnetos (another type of electrical generator) is driven directly by the engine, so that whenever the engine is turning, you are assured that the spark plugs will fire. There are two plugs in each cylinder, one fired by the left magneto and one fired by the right. This arrangement promotes safety and efficiency, and the engine will run quite nicely on either set.

POWERPLANT LIMITATIONS

Each engine manufacturer is required by law to provide a set of operating limitations for each engine model, and the FARs require you to observe them. One of the most important limitations is fuel grade, which refers to the fuel's ability to

Airplane Engines: Principles and Operation

resist detonation, known as "knocking" or "pinging" in an automobile engine. Detonation (explosion of the fuel-air mixture inside the engine instead of rapid burning) cannot be heard above the noise of an airplane engine, and if it continues for only a short period of time, irreparable damage will probably result.

Most light airplanes today operate on 100LL gasoline, the most prevalent grade available; "100" describes the grade, and "LL" refers to low lead, indicating the antidetonation additive, tetra-ethyl lead, has been reduced to the minimum. Small airplane engines were originally designed to run on 80-grade avgas, and it is the preferred fuel for these engines; unfortunately, availability of 80-grade is a problem in some areas.

There's a very important rule to be followed when selecting fuel for an airplane engine: you may safely use a grade higher in number than the one specified for your engine, *but never a grade lower.* (As of this writing, there are a few airplane engines in which automobile gas may be used legally.) Check the fuel limitations for your engine very carefully, and don't exceed them.

Additional engine limitations deal with maximum temperatures and pressures, as well as power settings; the latter may take the form of maximum RPM permissible, or limiting combinations of manifold pressure and RPM for those engines with controllable-pitch propellers.

The Federal Aviation Regulations make it unlawful for anyone to operate an airplane in violation of its prescribed limitations. This doesn't mean that you will be placed under arrest for running your engine too fast, but the law is on the book for safety's sake. Any piece of machinery which is operated beyond its design limits may come apart sooner or later, and the coming apart could be disastrous in an airplane.

7

Preflight and Engine Start

THERE ARE two very clear mandates in the Federal Aviation Regulations regarding your responsibility for preflight actions: first, you are required to familiarize yourself with all available information concerning the flight you're about to undertake; and second, you are responsible for determining whether the airplane is in condition for safe flight.

The FARs require more attention to detail when you are planning a cross-country trip, but this chapter is limited to the operational aspects of inspecting your airplane before any flight, and a discussion of engine-starting techniques.

THE PREFLIGHT INSPECTION

"Kicking the tires" just won't do it. When you set out to preflight your airplane, there are a number of critical items

Preflight and Engine Start

to be checked—enough of them to require some sort of organized procedure. The Pilot's Operating Handbook contains a recommended preflight routine; if none is available, your Flight Instructor can help you develop a checklist of your own.

Begin by taking an overall look at the airplane as you approach it. This "big picture" may uncover gross discrepancies, and might even save time; if you notice a flat tire, or a pool of oil under the engine, you might as well call the mechanic—you've got problems.

An interior check should precede your walk-around inspection; make sure that the magneto (ignition) switches are OFF, the throttle is CLOSED, and the mixture control is OFF. This is to preclude the possibility of the engine coming to life while you're moving around within striking distance of the propeller.

Check for loose objects in the cabin (they can become very distracting in flight, and have the potential for jamming controls). When you will be flying solo, it's a good idea to buckle the unused seat belts and shoulder harnesses, and take up all the slack, so that they can't possibly get tangled in the controls. This is especially important in tandem-seat airplanes.

Most pilots begin the exterior inspection right outside the cockpit door, and it doesn't make much difference which way you progress from that point. It is, however, very important to conduct your preflight the same way every single time, so that you develop a pattern for the inspection. A consistent, thorough preflight will cause discrepancies to stand out like sore thumbs; a hit-or-miss inspection almost guarantees that you'll miss something critical sooner or later.

You should take note of the general condition of the airplane, looking for loose parts, dings and dents, leaks (check for stains on the hangar floor or on the ground), and anything else that might indicate an abnormal situation. Move all the control surfaces to make sure there's nothing binding or sticking, and listen carefully all the while; you can often hear the squeak of a too-tight fitting or the rattle of a loose control cable. And, since the next time an airplane's controls are rigged backwards won't be the first time, always make absolutely certain that the control wheel or stick and rudder pedals are moving in concert with the control surfaces themselves; this will be checked again before takeoff, but it's a vital part of your preflight. (A thorough control check also precludes the possibility of leaving control locks installed.)

The single most important item in the preflight inspection process is the fuel check. There is no excuse for a pilot—any pilot—not knowing how much fuel is on board his airplane; it's a matter of looking into the tank(s) for visual confirmation, even using a dipstick if necessary. You must never trust someone else's word ("I just flew it around the pattern, there's a couple of hours left"), you must never trust the fuel gauges, you must see and confirm for yourself. Remember that no matter how short the flight, there must be a reserve of at least thirty minutes' worth of fuel on board, and you'll be more comfortable with a full hour in reserve.

The fuel check must encompass quality as well as quantity. To facilitate this part of the preflight, all airplanes are equipped with quick-drain fittings at the lowest point in the system. Draw a sample into a clear container, and check for two things: water or other contaminants, and the proper fuel grade. Aviation gasolines contain dye to identify the grade; 80-grade is red, 100LL is blue, and 100 "regular" is green.

It's almost impossible to conduct a too-thorough preflight inspection. In addition to fulfilling your responsibility to determine whether the airplane is in proper condition for safe flight, a good preflight uncovers the little problems before they grow into large, expensive ones.

And, by the way, don't forget to untie the airplane; it's very embarrassing to try to taxi with the tail still fastened to the airport!

ENGINE-STARTING TECHNIQUES

An airplane engine must be rotated in order for the combustion process to begin, so some sort of auxiliary power is required for starting. Depending on the age of your airplane and the equipment installed, this power source may be an electric starter or an "Armstrong" starter, which uses muscle power.

Electrical starter or not, there are three elements that must be present in the proper proportions if you are to get the engine running: air, fuel, and ignition. The spark plugs will begin firing as soon as the engine begins to turn (many engines are equipped with a "shower of sparks" system which provides a hotter, stronger spark for starting), airflow into the engine is adjusted by means of the throttle, so all that remains to be supplied is fuel.

In nearly all light-airplane engines, a small plunger-type fuel pump (primer) is mounted on the instrument panel to provide the initial charge of fuel, which is squirted directly into the intake manifold. When the engine is cranked, the fuel is drawn into the cylinders; if the fuel-air mixture is combustible, the engine will start and continue running on its own.

It is impossible to know exactly how much fuel should be used in the priming process, and you'll need to discuss this in

Recreational Flying

depth with your Flight Instructor. Every engine is a little bit different. Suffice it to say that the colder the engine and the outside air, the more prime is required to get the engine running.

Experience will also teach you where the throttle should be positioned for a proper start. You don't want the throttle wide open (a "full power" start is noisy, hard on the engine, and blasts the folks behind you), but it must admit enough air to mix properly with the fuel being introduced. For each starting situation, there's a throttle setting that will work—practice and experimentation will pay off.

When the air (or the engine) is very hot, remember that even a small amount of fuel in the intake lines will vaporize readily, and may provide a mixture adequate for starting. When in doubt, crank the engine a couple of revolutions and see if it will start. Emphasis on "a couple of revolutions"—if an airplane engine doesn't start readily, continued cranking will probably not solve the problem, and will eventually ruin the starter or deplete the battery. Unless the proper combination of fuel and air is present, the engine will not start; cranking seldom helps.

In most situations, failure to start after just a few revolutions indicates the need for additional fuel, so try another shot of prime (just one!) and crank again. When there's enough fuel in the mixture (it may take a couple of tries), the engine will catch up and run. If, after a couple of times through the priming process, the engine refuses to start, the problem could be too much fuel; you now have a flooded engine. In this case, a popular remedy is to crank the engine with the throttle wide open, but be ready to close it rapidly just as soon as the engine fires. (The open throttle admits more air, ventilating the engine, and leaning the mixture toward a more combustible ratio.)

Preflight and Engine Start

The techniques for hand-propping an airplane engine are the same in principle, but it's obviously more important for the pilot to set up the engine controls properly the first time. Although electric starters can crank the engine for many consecutive revolutions, it's "one turn at a time" with a human supplying the power. You will become very good at judging the prime and throttle setting when working with a hand-propped airplane—especially if you have to share the propping chores!

Starting any airplane by hand is an *extremely dangerous procedure*, and doing it by yourself merely heightens the hazard. There's a lot of technique involved and hand-propping is not something you can safely teach yourself—your first mistake may be your last. Find someone who knows how, and be sure you understand the entire process. Never attempt a hand-propped start unless the airplane is tied down, or a knowledgeable person (preferably another pilot) is seated at the controls.

No matter how you start the engine, you owe a warning to anyone who might be close by that you're about to fire up. When the prestart checklist is complete, when all that's left to do is turn the key or swing the prop, holler (yes, *holler!*) "CLEAR!" so that everyone within earshot will know that the rotary guillotine on your airplane is about to be placed in motion. There's a tongue-in-cheek definition which applies here: a microsecond is the period of time that elapses between a pilot's call of "Clear!" and the turning of the propeller blades. 'Nuff said?

Ground Operations: Taxiing and Before-Takeoff Procedures

UNLIKE other vehicles, an airplane is a creature of two environments—ground and air—and pilots need to become adept at handling their machines in both situations. This chapter will point out some procedures and techniques for getting your airplane to and from the runway (taxiing), and to help you develop an easily remembered, effective before-takeoff check procedure.

TAXI TECHNIQUE

For all practical purposes, an airplane is moved about on the ground by the same force that moves it through the air—

Ground Operations: Taxiing and Before-Takeoff Procedures

thrust from the propeller. As you open the throttle and cause the prop to rotate faster, it develops more and more thrust, until at some point, the airplane's inertia is overcome and it begins to roll across the ground. Open the throttle more, and the airplane will roll faster; keep this up, and of course the airplane will eventually reach flying speed.

But that's not the objective here. After the engine is started and all the preparations for moving out are complete, slowly open the throttle (squeeze, don't push!) until the airplane barely begins to move. It will always require more power to get the airplane moving than to keep it moving, so be prepared to reduce power as soon as the inertia is overcome.

Brakes should be checked before the airplane moves more than a few feet; if there's no stopping power available, it's better to discover it now.

Once under way, adjust the throttle to maintain a reasonable taxi speed, one fast enough to minimize the time between ramp and runway, but not so fast that the airplane cannot be brought quickly to a stop should the need arise. You'll soon discover a comfortable taxi speed—if it feels like you're moving too fast, you probably are. Resist the temptation to control taxi speed with the brakes; airplane brakes are not heavy-duty appliances, they're rather small, and they will overheat quickly, reducing their effectiveness and leaving you brakeless when you need them the most. Make small power adjustments to keep taxi speed where you want it.

Most of the airplanes you'll encounter as a Recreational Pilot will be equipped with a ground-steering system operated by the rudder pedals. The nose-wheel or tail-wheel turns to left or right as you push on the appropriate rudder pedal, with adequate steering authority for all but the sharpest turns. When you must turn the airplane in a very short radius on a paved surface, the brakes may be used independently, but be careful to keep the inside wheel turning; otherwise, a lot of expensive rubber will be scrubbed off the tire. (On grass or dirt, no problem.)

While good technique precludes the use of brakes during taxi, they will normally be required to bring the airplane to a complete stop. Use slow, steadily increasing pressure on the brake pedals (squeeze more on one than the other if necessary to keep moving straight ahead) until the airplane is nearly stopped, then ease up on the brakes. Remember that as power should be reduced after inertia is overcome when starting to taxi, brake pedal pressure should be reduced just prior to stopping; saves wear and tear on passengers' necks, and cuts down the chances of your being sued for whiplash injuries! Taxi smoothly; it's another mark of a good pilot.

You can bet that the captain of an airliner isn't very concerned about the effects of wind on his airplane while taxiing, since it takes a lot of wind to disturb a machine that size. The pilot of a light airplane has the harder job in this regard; not only will a tailwind increase taxi speed and a headwind slow you down (in either case, make the necessary adjustment in power setting), but wind blowing from either side will create a strong weathervane effect. You'll see this in the form of yaw, as the nose moves to the left or right on the horizon; take care of it with rudder pressure, just as you would in flight, and remember that you will undoubtedly need to hold some of that pressure all the time you are taxiing in a crosswind.

The Wright brothers found out the hard way what wind can do to a light airplane on the ground; the 1903 *Flyer* was completely wrecked following the historic first flight when a gust picked up one wing and proceeded to roll the airplane up in a ball. There's a lot of surface for the wind to work on,

Ground Operations: Taxiing and Before-Takeoff Procedures

but there are also controls with which you can "fly" the airplane while taxiing. When the wind is blowing from straight ahead, keep the wheel or stick all the way back, using the aerodynamic force of the elevator to hold the tail down. When taxiing in a following wind (from behind), the wheel or stick should be held forward so that the wind will push down on the elevator and prevent a nose-over. The lighter the airplane, the more important this becomes.

When the wind is blowing from the side, ailerons come into play. Your airplane's center of gravity is relatively high, and the main wheels are relatively close together, an ideal setup for an upset when a side force (crosswind) is introduced. Use this general rule when taxiing in a crosswind to reduce the possibility of overturning: when the wind is blowing from ahead and either side, set the controls as if to *climb and turn into the wind;* when the wind is blowing from behind and either side, set the controls as if to *dive and turn away from the wind.* And if the airplane appears ready to roll over even with full control deflection, let it turn into the wind, shut off the engine, and wait for the wind to subside.

Especially when operating high-wing airplanes, be very cautious when turning into the wind. In this situation, the push of the wind on either side of the airplane and on the underside of the upward wing is aided and abetted by the centrifugal force of the turn. Should these three forces in combination be strong enough, over you go—at the least, you'll scrape a wingtip. A good general rule is to slow down when the wind speeds up.

Taxiing on unprepared surfaces, such as grass or gravel, requires some additional considerations, particularly if there's any doubt about the quality of the surface (rough spots, loose gravel, or debris). Your primary concerns should be rough or soft ground (slow down to ease the bumps and to reduce the potential of nosing over) and propeller damage (keep the wheel or stick all the way back to gain all possible prop clearance from the ground, and don't use any more power than is absolutely necessary; the prop will suck up debris and rocks, causing expensive damage).

BEFORE-TAKEOFF PROCEDURES

Assuming that your preflight inspection uncovered nothing out of the ordinary, that the engine start was routine, and that you've arrived at the head of the runway with all the parts of the airplane intact, it's time to go through the before-takeoff check to make sure that all is in order. This is your last chance to conform with the regulation requiring you to determine the basic airworthiness of the airplane.

For pilots of larger, more complex machines, there's a host of items to be checked and tested; but the Recreational Pilot can develop an easily memorized before-takeoff checklist that will insure the essential elements are in proper operating condition. You'll find a number of memory-joggers for this purpose, and you may want to make up your own; nothing wrong with that—the important thing is to have some kind of checklist that is used *every single time*, and which guarantees that those items critical to safe flight are checked.

CIFFTRS is such a checklist; pronounced "sifters," it's a mnemonic, made up of the first letters of these words: Controls, Instruments, Fuel, Flaps, Trim, Runup, Seat belts. Here's the procedure for each item in CIFFTRS:

Ground Operations: Taxiing and Before-Takeoff Procedures

CONTROLS   Move the control wheel or stick to the right, look to the right, and check that the right aileron is up. Now move the wheel or stick to the left, check that the right aileron is down, then look to the left and check that the left aileron is up. Then move the wheel or stick to the right and check that the left aileron is down.

Turn in your seat so that you can see the tail surfaces, and move the elevator and rudder controls through their full travel, checking for proper movement of the surfaces. It's a good idea to check all the "corners" of control movement; with the wheel or stick full forward, move through the full travel of aileron control, then do the same in the full up-elevator position. If there is any binding in the control cables, you'll likely discover it here.

INSTRUMENTS   There's not much to check in this regard on a very simple airplane, but set up a routine; start in one corner of the instrument panel and proceed with an orderly visual check of whatever instruments are installed. Airspeed indicator (should be reading zero), altimeter (set to field elevation), magnetic compass (should agree with a known direction), fuel gauges, and engine instruments are the critical ones.

FUEL   Make sure that the fuel valve is ON (or the proper tank selected), and that there is sufficient gasoline aboard for the upcoming flight. This is also a good time to check that the mixture control is set to the full RICH position.

FLAPS   If you checked the wing flaps through their entire travel during the preflight inspection, you need only check them in the UP position at this time. If you'd rather make the check now (or check them again), be sure that the flaps move to the desired position, and that they move together.

TRIM   After a few cut-and-try exercises, you will learn where the elevator trim should be set for takeoff (and rudder trim, if installed). It's a good idea to run the trim through its entire travel to check for binding and proper operation (turn and look at the tab to make sure). In tandem-seat airplanes, your first solo flight will probably include a surprise, as you discover that the elevator trim setting which worked well when your instructor was in the back seat is no longer valid!

RUNUP   Just because the engine started promptly and has run smoothly up to this point doesn't mean that it will accept the demands of takeoff and flight. The manufacturer has provided explicit instructions for the engine power check (runup), and they must be followed to the letter. In nearly every case, you will check for smooth operation at a prescribed RPM (it's a fairly high power setting, so hold the wheel or stick full back, and hold whatever brake pressure is required to keep the airplane from moving), then operate the mag switch to check for smoothness and RPM drop when each set of spark plugs is cut out.

Although smooth running is probably more important than the amount of mag drop, you must begin the check at the prescribed RPM and note the amount of RPM decrease; an engine that loses more than the allowable number of RPM is in violation of its operation limitations, and may not be flown until the problem is resolved.

Next item in the runup procedure is a check of the carburetor heat system. When you operate the carburetor heat control, warm air (heated by passing through a heat exchanger on the exhaust system) is introduced into the carburetor for the purpose of preventing or eliminating the ice that often forms there. The warm air is considerably less dense, reduces the power output of the engine, and should therefore show up as a decrease in power (drop in RPM).

If carburetor ice was indeed present before the check, engine speed will drop perhaps several hundred RPM as the ice is melted and drawn through the engine, then will come back to a value slightly less than the no-heat setting, and should return to the original value when the heat control is moved back to

Ground Operations: Taxiing and Before-Takeoff Procedures

COLD. (By the way, this check is just as valid in flight, and should be performed whenever carburetor icing is suspected.)

SEAT BELTS Now that the machine has been thoroughly checked for flight, turn your attention to the people. Seat belts must be fastened for takeoff and landing (there's no good reason why they should ever be unfastened in flight), and you have a legal responsibility to make sure that your passenger is so informed. Loose, twisted belts are less effective, so make sure that all of them are snug, straight, and flat. If shoulder harnesses are installed on your airplane, *use them*—they save lives and reduce injuries.

So much for CIFFTRS. But there's one more thing you must do before leaving the ground—be absolutely certain that the runway is all yours. Before taxiing onto the runway, turn the airplane (make a complete circle, if you have to) so that you can confirm visually the absence of other airplanes on approach for landing. Remember that some pilots make up their own traffic pattern rules.

Training Operations

THE AIRMAN CERTIFICATION REGULATIONS set out a long list of operations in which Recreational Pilots must be trained. Even though it's unlikely that you'll be asked to demonstrate all of these on your checkride, the fact remains that your instructor is required to make you familiar with each and every one.

These operations are the bases on which you can build your proficiency for safe utilization of the airplane. In this chapter, we'll discuss the required in-flight operations only; takeoffs and landings are dealt with elsewhere. Keep in mind that the primary objectives of these operations are to teach proper stall recognition and recovery, and to teach traffic pattern procedures in preparation for the takeoff/landing phase of your training.

NORMAL CRUISE

Although your airplane is capable of cruising (flying level) at a wide range of airspeeds, there's a "normal" speed which

Training Operations

satisfies most training and cross-country requirements quite nicely. This speed usually results from a power setting of approximately 65 percent of the engine's rated horsepower (there's a chart or table that provides the RPM or RPM–manifold pressure numbers to set this up), and is also very close to the speed for which the vertical stabilizer is offset to counteract the yaw from P force.

Flying at the normal cruise speed is therefore comfortable (in terms of required control input), well above the stall speed, provides adequate forward visibility, good control feel, and adequate control response for training operations. As the most fundamental of all flight operations, normal cruise needs to be practiced until you can establish and maintain it well, noticing and responding to the various visual cues—wings level, no yaw, pitch attitude constant.

FLIGHT AT SLOWER-THAN-NORMAL AIRSPEED

This operation is also known as "slow flight," and is usually not defined in specific airspeed terms. Assuming that you will maintain a constant altitude throughout this exercise, it follows that any reduction in power will result in a lower airspeed; the amount is up to you.

Slow flight may be useful for observing things on the ground, or perhaps for adjusting your spacing behind another airplane in a traffic pattern.

Whatever the speed, there will be a difference in the flight controls; they will feel softer and the airplane's responses to your inputs will become more sluggish as airspeed drops. You will be operating closer to the stall speed, and some of the indications of aerodynamic stall will begin to appear (more on that later). The nose will ride higher on the horizon, since a higher angle of attack is required to provide adequate lift at the reduced speed. Yaw to the left will become more noticeable, and as airspeed is reduced, yaw control becomes critical. Use whatever rudder pressure is required to prevent yaw.

To recover from slow flight, add full power, adjust the pitch attitude to the normal cruise position on the horizon, and when airspeed builds to the cruise value, reduce power to the normal setting (preventing yaw with rudder pressure all the time, of course).

FLIGHT AT
MINIMUM CONTROLLABLE AIRSPEED (MCA)

This is strictly a training operation; there's no useful application in everyday flying, because you must pay too much attention to flying the airplane. Unlike slow flight, MCA has an airspeed definition: it's the lowest possible speed at which the airplane can be flown without a stall occurring. MCA is particularly valuable as a training operation because your perceptions of an impending stall become the "airspeed indicator." Properly done, MCA is truly flying on the very edge of a stall; it helps you develop a fine touch on the controls, and hones your attitude-flying skills.

To begin this operation, clear the area with at least two 90-degree turns, making certain that you're not sharing the airspace with others in close proximity. As you roll out of the second turn, reduce power, maintain altitude (higher pitch attitude required), and allow airspeed to decrease to a value

Training Operations

perhaps 10 to 20 MPH above stall speed. Note the pitch attitude, maintain it, and add just enough power to maintain altitude.

When you are able to hold both airspeed and altitude constant (with pitch attitude and power, respectively), begin to slow down by simultaneously raising the nose a bit and reducing power slightly. As you proceed deeper into the low-airspeed corner of your airplane's performance envelope, there will be a remarkable increase in the amount of rudder pressure required to prevent yaw.

After a series of small changes in airspeed (don't hurry, there's no prize for speed here) you'll notice that the controls are getting sloppy, the airplane is beginning to buffet (shake), and if there's a stall warning system installed it will be sounding or flashing. If you slow down any more, the wing will stall—you have arrived at Minimum Controllable Airspeed. Don't expect to accomplish the very slowest speed on your first few tries, and don't be put down if a stall occurs; it's like seeing how close you can walk to the edge of a cliff in the darkness—you know you're there when you fall off!

Yaw control becomes critical in this operation. Whenever yaw is present, one wing is moving through the air slightly faster than the other, which means that the faster wing is developing more lift. Should a stall occur in this condition, you can expect a rolling motion because of the unequal lift, and if the problem is not corrected, a spin will most likely result. Use the rudder as required to prevent yaw.

Recovery from MCA is accomplished with a procedure nearly the same as that used for slow flight; add full power, adjust the pitch attitude to the climb position (to minimize loss of altitude during recovery), and when the airplane has achieved a normal climb condition, accomplish the level-off procedure.

Recreational Flying

If your airplane is equipped with wing flaps, they should be extended in order to accomplish MCA. Lower the flaps as you roll out of the first clearing turn (more power will be required to overcome the additional drag), and on recovery, retract the flaps slowly after the addition of full power and establishment of the climb attitude.

STALLS

The words with which an aerodynamic stall is defined are relatively unimportant, but your ability to recognize a stall, your understanding of an airplane's behavior when a stall occurs, and your ability to recover from a stall with minimum loss of altitude is extremely important. For that reason, all pilots must be competent in several types of stall entries and recoveries, for the sole purpose of learning to recognize and recover from them.

The training operations discussed in this section cover those conditions of flight in which stalls will most likely occur, and the recovery procedures are designed to get you flying again as rapidly as possible. As you train, don't strive to become good at doing stalls; rather, become good at recognizing and recovering from stalls. It's also a very good way to get familiar with your airplane, so that its behavior in an adverse situation won't be a complete surprise.

Although stalls are usually encountered in a low-airspeed condition, keep this fact firmly in mind: *a stall can occur in any attitude, at virtually any airspeed—whenever the wing's critical angle of attack is exceeded, it will stall.*

What follows is a scenario recommended for your progression through the training stalls, from the softest, "pussy-

Training Operations

cat" stall to the sharp-breaking, critical ones. It won't be mentioned every time, but make it a practice to begin all of your training-stall exercises at least 2,000 feet above the ground, and accomplish at least two 90-degree clearing turns prior to each stall, for safety's sake.

With the help and advice of your Flight Instructor, and with reference to the airplane handbook, develop a simple, standard stall-recovery procedure so that you'll know what to do when the time comes. Recovery from any stall must begin with a reduction of the angle of attack, either by relaxation of the back pressure on the control wheel or stick, or a definite forward push to get the wing flying again.

One very reliable, workable recovery procedure for light airplanes is to add full power, place the nose of the aircraft in its normal climb attitude and *hold it there*, retract the wing flaps (if installed and extended), and continue the climb.

In any event, get as good as you can be during training, then stay sharp with a stall-practice session every now and then so that recognition and recovery becomes almost a reflex action.

Power-off Stall, Straight Ahead

From normal cruising flight, close the throttle and establish a normal glide (the one you'll use for the landing approach) straight ahead, then raise the nose slowly to the horizon and hold it there. As the airplane slows down, take note of the changes in wind noise, control feel, and the requirement for an ever-increasing amount of right rudder. Maintain this pitch attitude until the control wheel or stick is all the way back and the nose drops (note the airspeed at which the stall occurs, if you can).

Power-on Stall, Straight Ahead

For your first experience with a power-on stall, choose a power setting somewhat less than that used for cruise; we'll work up to the high-power stalls later. Perhaps 1,500 RPM is a good compromise on most light planes; it will serve to acquaint you with the differences in behavior when power is introduced, but without the abnormal attitudes and aerodynamic reactions encountered in a full-power stall.

Enter the power-on stall in the same manner as the power-off exercise, but raise the nose a bit higher (otherwise, the airplane may be content to fly level at 1,500 RPM at a very low but unstalled airspeed). You'll notice a distinct need for considerable right rudder pressure, as airspeed drops off and the P force asserts itself. When the stall occurs (at a lower airspeed than power-off), there will be a sharper "break" (dropping of the nose), and the airplane will exhibit a very strong tendency to roll and yaw to the left. Prevent yaw, prevent roll, and proceed with the recovery procedure.

Power-off Stall in Turning Flight

When an airplane stalls in a well-coordinated turn (that is, a turn with no yaw present), the reaction is very similar to a stall in straight flight. However, even a small amount of yaw will cause one wing to develop more lift than the other, and the airplane will begin to roll when the stall occurs. Carried to extreme, this will produce a spin; use whatever rudder pressure is necessary to prevent yaw.

Set up a normal glide straight ahead, then raise the nose smoothly to the horizon and roll into a turn to the left (a 30-degree bank is preferred). As soon as the airplane begins to turn, notice the rate at which the nose is sweeping across

Training Operations

the horizon, and maintain this turn rate with rudder pressure. The bank angle will tend to increase as the airplane slows down, but your job is to keep the 30-degrees constant until the stall occurs.

The usual stall indications and warnings will appear at a slightly higher airspeed because of centrifugal force in the turn. When the stall occurs, release back pressure on the wheel or stick, stop any yaw with the rudder, level the wings with aileron pressure, then go through the power and pitch exercise to regain normal flight conditions.

Be prepared to use more control travel than was required in the straight-ahead stall, because you must reestablish straight flight in order to recover from a turning stall with the least possible altitude loss. Use whatever control movement is required to get the job done.

Do this stall again to the right, and practice power-off turning stalls until you are confident of your ability to be the complete master of your airplane. (You'll notice a very definite difference in airplane behavior when comparing left- and right-turning stalls; nearly all airplanes show a strong tendency to yaw and roll to the left when stalled in a left turn—propeller forces are responsible—and a resistance to roll and yaw when turned to the right.)

Power-on Stall in Turning Flight

In this exercise, the aerodynamic forces really begin to make themselves felt. Entry and pitch attitude are the same as the power-on stall straight ahead, except that a medium angle of bank (20 to 30 degrees) is introduced as soon as the pitch attitude is established. Maintain the rate of turn with the rudder, maintain the angle of bank with aileron pressure, and go through the standard recovery procedure when the stall occurs.

The control inputs to maintain pitch, bank, and turn rate will not be noticeably greater than in the power-off stall, and you'll need to be more aggressive with the controls on recovery. The differences between left- and right-turning stalls will also be very evident.

Because of the additional forces generated by the propeller, it is even more important to prevent yaw in low-airspeed turns, when you are knocking on the door of a stall.

Approach-Landing Stall Series

This is the first of several training exercises that are intended to acquaint you with the stall characteristics of your airplane in conditions very much like those found in actual practice. The approach-landing stall series simulates the turn from base leg to final approach in the traffic pattern; its importance lies in training you to recognize and recover from an impending stall in a glide very close to the ground. The series is accomplished by doing a stall straight ahead, then a turning stall to the left, followed by a turning stall to the right.

Enter a normal glide straight ahead (if your airplane is equipped with wing flaps, extend them fully), then raise the nose to the landing attitude (use whatever your Flight Instructor has pointed out for visual cues), and recover when the stall occurs. Repeat the exercise in a left turn, then a right turn. When you are sufficiently confident, take note of the altitude when the stall occurs, and again when you have established a normal climb. Minimum altitude loss is the bottom line of this exercise.

Training Operations

Takeoff-Departure Stall Series

The second series of training stalls sets up the flight condition you would encounter very shortly after takeoff: the engine developing full power, nose relatively high, and airspeed relatively low. The high power setting guarantees a great deal of P force at work, and when the airplane stalls in this condition, a spin is virtually certain unless you take prompt, proper, and assertive action on the controls.

After your clearing turns, close the throttle, maintain altitude, and let the airspeed decrease to a value slightly less than that used for normal climb. Then simultaneously add full power and raise the nose to an attitude considerably higher than that used for the other stalls, and hold it there until the stall occurs.

Yaw and roll forces will be very strong (you'll need a lot of aileron and rudder pressure to make the airplane behave!), and the break will be rather sharp when the wing stalls. Release the back pressure, and smoothly reestablish the normal climb attitude (power is already set).

Repeat the stall in turns, in both directions. The high power setting aids in recovery with minimum altitude loss, but it also provides some very interesting aerodynamic reactions at the stall! Like never before, you must maintain proper coordination (no yaw) to keep a spin at arm's length; spins are not all bad, but there's seldom enough room between you and the ground for recovery if a spin is entered shortly after takeoff.

Accelerated Stalls

This is a demonstration-only operation for the Recreational Pilot, since it can easily place you in a condition of flight for which you are not prepared. The objective of an accelerated stall is to prove beyond a doubt that your airplane will stall at an airspeed considerably higher than the value quoted in the handbook.

Your Flight Instructor will clear the area, reduce power, and roll into a rather steeply banked turn, then haul back on the wheel or stick to increase the centrifugal force significantly and suddenly (you will also feel the G force on your body). The rapid increase in effective weight causes the airplane to "mush" through the air somewhat, the angle of attack is rapidly increased, and the stalling angle is reached at a much higher airspeed. The break will be rather sharp, with high roll and yaw moments.

If you learn nothing else from the demonstration of an accelerated stall, learn that you must be aware of the impending consequences whenever you feel the G loads building up; reduce the angle of bank, add power, lower the nose, whatever it takes to fly safely out of that condition—and *prevent yaw!*

SPINS

Recreational Pilots need to take special notice of spins. They've not been required in flight training programs for a number of years, primarily because contemporary training airplanes have been designed with spin-resistant characteristics. But many of the older airplanes that will be flown by Recreational Pilots were built in "the old school," which insisted that pilots be familiar with spins. As a result, these older airplanes will spin *very easily*. (There's a benefit here, however; an airplane that can be put into a spin without

Training Operations

much effort will also fly itself out of that spin if left alone.)

A spinning airplane is one that has been stalled in an asymmetric lift condition, with one wing producing more lift than the other. This of course generates a rolling movement, which, when combined with the poststall nose-low attitude, puts the airplane in an altitude-eating spiral. Airspeed will be very low because the airplane remains in a stalled condition, but vertical speed builds up to a rather frightening value; when recovery takes place, the airplane is headed straight down and airspeed increases rapidly.

You may find that spins are a lot of fun, and there's no doubt that a pilot who is not intimidated by spinning and who seeks training in the proper recovery procedures will not be surprised and helpless should he get into one inadvertently. *WARNING: Spins are considered a mild form of aerobatics, and the regulations require approved parachutes for all occupants when an airplane being flown by a Recreational Pilot is spun intentionally.*

GROUND REFERENCE MANEUVERS

Like a good sailor, a good pilot must be aware of the wind and its effects, and be able to counter those effects when they are adverse to the operation being conducted. The ground reference maneuvers that follow are intended to develop an awareness of the wind, and to acclimate you to flight operations relatively close to the ground. This is preparation for traffic pattern work.

All of these operations should be conducted at not more than 800 feet above the ground (typical traffic pattern altitude), and, of course, not less than 500 feet AGL for safety's sake and observance of regulations.

Tracking a Straight Line

In aviation language, a "course" is the line to be flown between two points, a "heading" is the direction in which the airplane is pointed, and the "track" is the actual path across the ground. When the wind is blowing from straight ahead or directly behind, course and heading will be the same; but when there's a crosswind present, the airplane must be turned somewhat into the wind in order to keep from being blown off course.

Since you must walk before you can run, your first exercise in this regard should be tracking a straight line. Pick out a fairly long, straight stretch of highway, railroad, or shoreline, and fly parallel to the line in both directions. You'll soon discover how much heading change (wind correction, or drift angle) is required to accomplish the desired track. Train yourself to use a visual reference on the airplane structure (a tire, a wing strut, a spot on the leading edge of the wing) so that you can see the drift and correct it.

Rectangular Pattern

Once the straight-line technique is mastered, move on to flying a simulated traffic pattern—a rectangle around a large field, a land section, intersecting highways, or the like. This is a further development of your drift-correction skills, and involves changing heading in both directions, and at the completion of each 90-degree turn. Why the importance of flying a precise rectangle in the traffic pattern? So that you will arrive at the same position each time in preparation for beginning the glide to landing; you'll establish a "groove" that will make it much easier to judge the quality of your approach and make adjustments.

S-Turns across a Road

This operation carries the wind-correction technique a bit farther, requiring you to adjust heading continuously as you fly in a series of half-circles. It will also help you to begin flying the airplane without paying much attention to what's happening inside.

It's important to choose a road (a railroad track or straight section of shoreline will do just as well) that lies perpendicular to the prevailing wind. The objective is to fly a continuing series of half-circles that begin and end over the road with wings level, with all the half-circles the same size.

This is one time when a demonstration of how not to do something makes it easier to understand the correct procedure. If you should approach the road on a perpendicular track with the wind behind you, and roll into a 30-degree bank as you cross the road, the wind would displace the airplane a considerable distance before the turn is completed—you'd need to roll out and fly straight ahead to get back to the road, and that doesn't satisfy the objectives. Conversely, if you approach the road from the other direction (into the wind) and roll into the same turn, the wind will retard your progress while the airplane is turning, and you would cross over the road before the turn is complete.

The proper way to solve the problem is to change the rate of turn by changing the angle of bank, so that the loop size will remain constant. Enter the S-turn again with the wind at your back, roll into the 30-degree bank, but about one-quarter of the way around, begin rolling out ever so slowly,

Recreational Flying

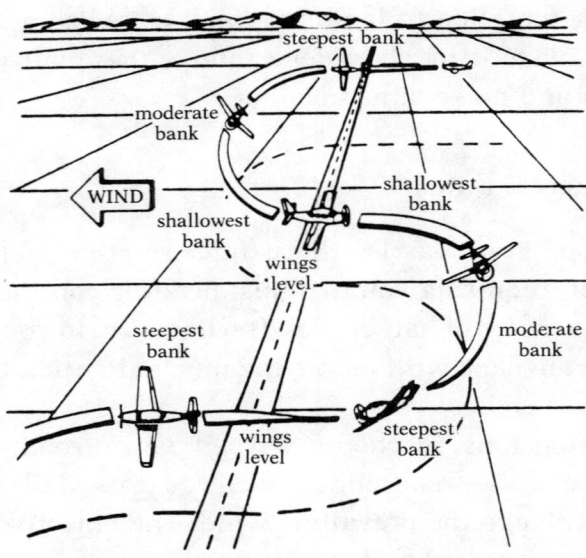

S-turns across a road

so that you can arrive back over the road with wings level, having completed a perfect half-circle.

The secret is to visualize the relationship of turn rate and ground track; adjust the angle of bank (it won't require much change!) so that your airplane flies in a consistent circular path. When approaching the road into the wind, you'll need to start off with a very shallow bank, and increase it gradually as you see the need.

When you can fly S-turns of reasonable consistency, maintaining altitude and airspeed (with only occasional glances inside to check those indicators), you should move on to the next ground-reference maneuver.

Turns about a Point

This operation accomplishes essentially the same purpose as S-turns, but takes up much less airspace, and uses

Training Operations

just one major reference point. Choose a highway or railroad intersection (preferably with a 90-degree crossing angle), or a lone tree in the middle of a large field—any prominent point with plenty of clear, flat space around it will do.

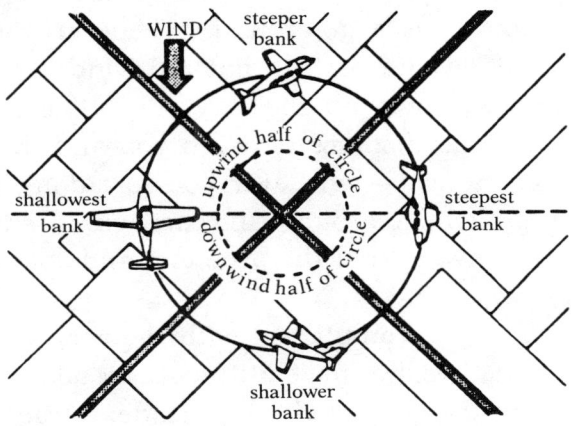

Turns about a point (viewed from above)

Enter this maneuver with the wind at your back, and when the point is directly abeam (off a wingtip), roll into a medium-banked turn. Put your S-turn skills to work, and adjust the angle of bank to fly a half-circle, and when the point is abeam the *other* wingtip, continue the turn to complete the circle. Turns about a point are nothing more than S-turns without the change in direction.

It's very helpful to pick out four references (houses, other intersections, trees) at the cardinal points on the circle you intend to fly, so that you can better judge the radius of the circle. In this exercise as well as S-turns, it's wise to start out with relatively large circles, and work up to the more demanding, small-radius patterns. You'll soon see that wind velocity also plays a part in determining how large your circles will be.

The Go-Around

There's a safety valve for nearly everything we do, and in flying, it's known as a go-around. Whenever you are approaching to land and realize that the wind is too strong, that you are not lined up the way you'd like to be, that there's not enough distance between you and the airplane ahead on the runway—whenever you just don't feel right about an approach, go around! In other words, enter a normal climb, get away from the problem, and come back for another try.

A go-around is accomplished with the very same procedure you learned earlier for stall recovery; add full power, put the nose in the normal climb attitude, retract the wing flaps if appropriate, and climb out of trouble.

Don't let pride get in your way; the smart pilots are the ones who recognize a nonstandard or abnormal situation and come back for another try, using the knowledge gained on the first effort to make the necessary adjustments. And whenever you go around because there is another plane on or near the runway, fly off to one side so as to keep the other fellow in sight.

When in doubt, go around!

10

Normal Takeoffs and Landings

THIS CHAPTER deals with perhaps the most important pilot operation of all, for no matter how good you may become at putting your airplane through its paces aloft, you still must get it off the ground and back on again smoothly and safely. And you'll be judged most harshly by passengers and fellow pilots alike on the quality of your takeoffs and landings.

In order to provide a logical sequence, and to give you something to build on as you progress to the more demanding operations, this discussion is limited to "normal" takeoffs and landings. These are the procedures you'll use most frequently, with little or no crosswind, no need for maximum performance, everything on the airplane working properly, and no emergency conditions to cope with.

Both power-off and partial-power approaches will be discussed, as well as techniques to be used with nosewheel and tailwheel airplanes. For the purposes of this chapter, we'll "stay in the pattern"—a closed circuit such as you will use when doing repetitive takeoffs and landings for practice.

Assuming that you've completed your before-takeoff checklist, and have assured yourself that the final approach is clear of other traffic, let's go!

TAKEOFF PROCEDURE

There are at least two distinct methods of beginning the takeoff roll: you may taxi onto the runway and come to a complete stop on the centerline, make sure everything (including the pilot) is in the proper state of readiness, then release the brakes and take off; or you may accomplish a "rolling" takeoff, in which the entire get ready–add power–line-up-and-go process takes place while the airplane is accelerating along a continuous, curved path from the runup spot to the centerline of the runway. Make it easy on yourself at the start; use the former procedure for at least your first few takeoffs.

In either case, you will feel the airplane come alive as power is applied. Specifically, P force will assert itself in the form of considerable yaw to the left; at this very low airspeed, P force is strong and rudder authority is relatively weak. In virtually all conditions, you will need to apply and hold some right rudder pressure throughout the takeoff roll. How much? Whatever it takes to keep the nose from moving to the left; prevent yaw, even if it requires full control travel. Open the throttle slowly and smoothly, and you'll be able to stay well ahead of the yaw problem.

Normal Takeoffs and Landings

Very shortly after the throttle is opened, there's enough air flowing over the elevators to bring them to life as well, and you will be able to establish the proper takeoff pitch attitude. In a tricycle-gear airplane, this means raising the nose to a predetermined visual relation with the horizon (your Flight Instructor will point this out very clearly); if you are flying a taildragger, you'll need to lower the nose a bit, just enough to get the tailwheel off the ground. In either case, it's important to establish the proper pitch attitude and *hold it* until the airplane lifts off the runway. A good normal takeoff is one in which the pilot is surprised to find he is airborne; don't pull the airplane off the ground . . . let it fly when it's ready.

Yaw control becomes even more important (and a little more involved) when the takeoff attitude is established, since you are now rolling on the main wheels only, and the airplane is free to pivot in response to the various forces at work. Make many small corrections with the rudder, rather than a few large ones; a pilot can fall behind in a hurry, and the takeoff roll becomes a weaving, tire-squealing experience. Remember that it will probably be necessary to hold some right rudder pressure throughout the takeoff roll because of P force, and therefore it's unlikely that the rudder pedals will remain lined up. The proper rudder-pedal position during takeoff is whatever results from preventing yaw.

When the airplane lifts off the ground, establish the normal-climb pitch attitude, again determined by placing some part of the airplane structure (the engine cowling, for example) on the horizon. As airspeed increases, the need for right-rudder pressure will drop somewhat, but there are precious few single-engine airplanes that don't require continued pedal pressure during a climb. Hold the wings level with aileron pressure, prevent yaw with the rudder, and you will be in completely coordinated flight—just what your Flight Instructor wants!

At an altitude of 500 feet above the ground, turn left or right 90 degrees (direction of turn depends on the traffic pattern in use at the airfield; most are left-hand patterns), and continue climbing to the downwind leg. That completes a normal takeoff.

TRAFFIC PATTERN

Here's where your training in rectangular patterns (correcting for wind drift while flying at a relatively low altitude and directing most of your attention outside the airplane) begins to pay off. A traffic pattern is simply another rectangular pattern exercise, with a takeoff or a landing added.

A proper traffic pattern serves two very important purposes at an uncontrolled airport: it organizes the flow and provides a reasonable anticipation of others' positions and actions, and it is the foundation upon which you can build a reservoir of visual cues to determine the corrections necessary for a good approach to the runway. In other words, when you have established a standard for your traffic patterns (same altitude, same airspeed, same track across the ground every time), you'll soon be able to make judgments regarding when to start the turns to base and final, the amount of power required to stay on the glide path, and so forth. If you don't know, you can't adjust!

Most patterns are flown either 800 or 1,000 feet above the ground, but there are airports at which the pattern altitude is higher, usually because of nearby residential areas, hospitals, towers, or other structures. Check local requirements

Normal Takeoffs and Landings

and practices when you fly to a strange airport; it's the pilot's responsibility to be aware of these things.

You should also find out if the airport you intend to use as a Recreational Pilot is also used by instrument pilots. Even though there's no control tower, there may well be a published instrument approach procedure for the field, in which case a conflict can arise as to which pilot—Recreational or Instrument—has the right of way. This is particularly important when the ceiling or visibility values are something less than "clear and a million," when both types of operations may be in progress. Both are legal (assuming a flight visibility of at least three miles and adequate cloud clearance for the Recreational Pilot), but both pilots also assume the responsibility for collision avoidance. When you've any concern for inbound instrument flights, be extra vigilant when your pattern takes you through the airspace that might also be used by your IFR brethren.

Since takeoff and landing performance is enhanced by a headwind, a normal traffic pattern is described in terms of its relationship to an upwind (into the wind) operation. The legs, or segments, of the pattern are known as upwind (takeoff), crosswind, downwind, base, and final approach.

The size of the pattern will depend on a number of factors, not the least of which is the size and speed of the airplane involved. For most of the airplanes used by Recreational Pilots, the downwind leg determines pattern size, and should be flown one-half to one mile from the runway. (In high-wing airplanes, it's convenient to locate yourself on downwind so that the runway appears halfway up the wing strut. There are equally usable visual references on other airplanes, and you should develop such a reference for your airplane.)

A very general rule of thumb for the vertical profile of a

Recreational Flying

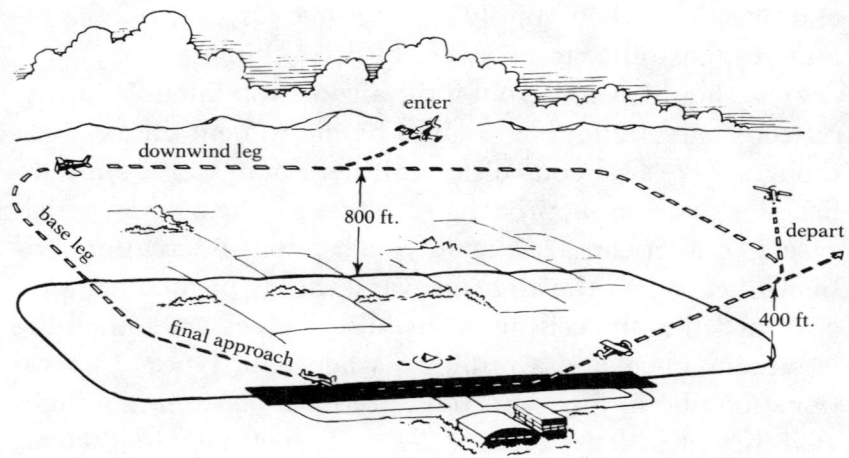

A typical standard airport traffic pattern

traffic pattern calls for you to gain one-half of the desired altitude while climbing straight out on the upwind leg, and continue the climb on the crosswind leg. On the other end of the pattern, plan to lose one-half of your altitude while flying the base leg, which places you on final at a safe altitude, with plenty of time to adjust the glide path for a good landing.

There's an extremely important reason for adhering to the accepted standards of altitude and pattern shape; if all pilots abide by the rules, each one will know where to expect and look for the other. In most midair collisions in the traffic pattern (which is where a large number of them occur, by the way), one or both of the pilots are not flying a standard pattern; it's pretty tough to avoid a collision when the other guy is not where you expect him! The safety aspect of a traffic pattern is null and void unless all pilots comply—especially when there is no radio communication, as in recreational flying.

Normal Takeoffs and Landings

PREPARATION FOR LANDING

Just as it was considered a good idea to use some sort of checklist to make sure that you had accomplished the critical items before takeoff, you should use a memory jogger prior to each landing. A simple, easily remembered checklist allows you to concentrate on flying, and be assured that all is in readiness for landing.

Again, there are a number of mnemonic acronyms (words manufactured with the first letters of words), and one of the most popular is GUMPS, with which you are reminded to check Gas, Undercarriage, Mixture, Prop, and Seat belts.

Recreational Pilot–type airplanes won't have retractable landing gear (undercarriage), and very few will have controllable-pitch propellers, but it's not a bad idea to form the habit of checking these items anyway, in case you move up to such complex airplanes later on. This is especially true with regard to flying a "retractable," where the embarrassment of landing with the wheels up is overwhelming (and expensive!).

The place to perform the GUMPS check (or whatever method you elect to make sure these critical things get done) is on the downwind leg, *every time*. Even if you stay in the pattern for landing practice and have gone around the circuit a dozen times, force yourself to go through GUMPS each time, just to cement the habit in your mind.

APPROACH AND LANDING PROCEDURE

Many of the airplanes available for Recreational Pilot training are very light, glide well, and are not equipped with wing flaps. These characteristics virtually demand a power-off, gliding approach in order to keep pattern size within reasonable limits; even a small amount of thrust will reduce the rate of descent remarkably, and the final approach leg will be very shallow and very long. We'll use the power-off, no-flap approach as the norm, and progress to the variations later.

It's important to arrive at a point directly opposite (abeam) the intended landing spot at the same altitude, the same airspeed, and the same distance from the runway each and every time. Not only does this force you to be in the right place with respect to others' anticipation of your position, it also provides a basis for judging the quality of your approaches and for making the appropriate corrections next time around.

From this position, apply carburetor heat (icing is much more likely when the engine is idling), close the throttle completely, and set up a glide at the airspeed recommended by the manufacturer. (No such information in your airplane's handbook? You'll need to rely on your Flight Instructor's knowledge and judgment, and a considerable amount of cut-and-try until you find a comfortable airspeed.) For the time being, we'll assume a no-wind condition; when the landing spot appears about 45 degrees behind you (on a line midway between wingtip and tail), turn to the base leg, descending all the while. You should plan to lose half of the pattern altitude just before turning onto the final approach leg.

Don't be upset if considerable practice is required to roll out precisely on the extended centerline of the runway; there's a lot of judgment involved, and it will take time to develop.

At this point in the approach, you should set up a visual reference to determine the probable touchdown point. Every

Normal Takeoffs and Landings

airplane windshield has spots on it (dirt, bugs, oil), so choose one (even if you have to imagine it!) in the line of sight from your eyes to the runway. This spot becomes a "gunsight" reference, and if you are careful not to change the pitch attitude of the airplane or raise or lower your head, the spot will indicate the probable touchdown point on the runway. If the spot appears to move toward the far end of the landing strip, you will surely overshoot; if the spot rests short of the runway, that's where you are bound. The ideal touchdown point for normal landings is within the first third of the landing surface.

Now comes the tough part. Down near the bottom of the approach, not very far above the runway, you must accomplish one of the most difficult tasks in the pilot's repertoire: the roundout and touchdown. Keep in mind the aerodynamics of this situation; you must transition from a glide to a power-off stall, and manage the energy in the airplane so that the stall occurs just as the wheels touch the runway.

The roundout consists of a gradual change in pitch attitude which, with no power applied, will result in a decrease

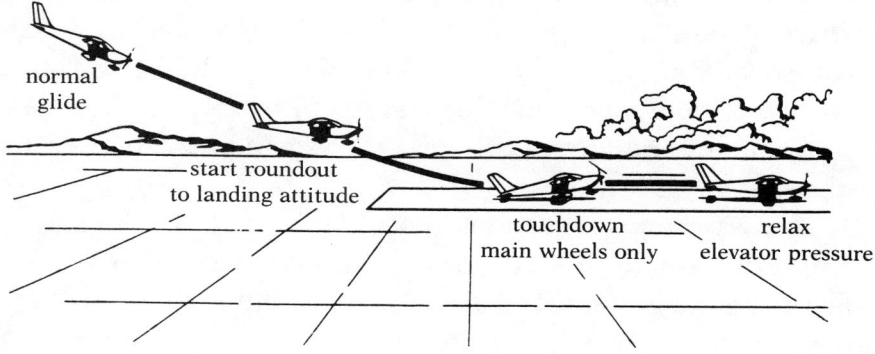

Normal landing sequence and procedure

in airspeed, an increase in drag, and obviously, an inevitable stall. If you increase pitch too rapidly or too soon, the airplane will stall well above the surface and fall to the ground—ungracefully, to say the least. If you are too slow, or wait too long to begin the roundout, the airplane will arrive on the runway in full flight, with potentially disastrous results.

The ideal touchdown is very gentle, at the lowest possible airspeed, so that the wings stall (give up all their lift) at just about the same moment the wheels touch the ground. This can be assured by establishing a visual reference that places the airplane in the same attitude used for takeoff: with tricycle landing gear, this will result in ground contact on the main wheels only; for the taildraggers, a perfect three-point landing, both main wheels and the tailwheel touching at the same time.

You should experiment with different visual references during that frustrating time of flight training when you are trying to "find the runway." There's a lot of skill involved in combining the timing and the rate of pitch attitude change, controlling yaw and bank, all the while managing the energy so that the stall occurs at the right time. Be prepared to muff a lot of roundouts in the learning; you'll fly through your share of hard landings because you waited too long or flared too soon, you'll learn that a bit of power will cushion an early roundout, and that you can recover from a bad bounce by going around for another try. That's what learning is all about!

It's very important to maintain the touchdown pitch attitude (nose on the horizon, or whatever) until the airplane is firmly on the ground. The airplane will exhibit a strong tendency to pitch down, and you must hold whatever back pressure is required to keep the nose where you know it must be.

Normal Takeoffs and Landings

Of course, yaw control is no less important here than on takeoff.

If you are learning in a side-by-side airplane, be sure to choose a visual reference point directly ahead to maintain directional control. Most pilots try to put the center of the airplane cowling on the centerline of the runway, which automatically builds in a yaw to the left. Since you are seated to the left of the airplane's centerline, realize that your visual reference for straight travel must also be straight ahead. Sight along a line slightly to the left of the runway centerline, and the airplane will go where you want it to.

Following touchdown, hold the control wheel or stick all the way back; in tricycle-gear airplanes, this guarantees that you'll roll on the main wheels only until it's safe to put the nosewheel on the ground. (This technique generates a lot of drag to help get stopped without use of the brakes, and also protects the relatively fragile nosewheel from unnecessary loads.) Full back-stick is absolutely required after touchdown for taildraggers, since steering effectiveness is greatly increased with that additional weight on the tailwheel. The center-of-gravity location of a tailwheel airplane contributes to a rather unstable directional control situation on the ground and you'll want all the steering authority you can muster!

As a courtesy to those pilots who may be right behind you in the landing pattern, turn off the runway at the first available taxiway—but don't attempt to turn at a speed that might invite the problems of centrifugal force or gusty winds. When in doubt, continue to the end, and if the other guy is so close that he has to go around, so be it.

Once well clear of the runway, take care of those items that should be turned off or repositioned; if your airplane is sufficiently complex, you may want to use an after-landing checklist to make sure everything gets accomplished. And remember that a flight is not over until the airplane is once again tied to the ground; keep "flying" (maintain complete control) until you are back in the chocks.

PATTERN ADJUSTMENTS

It won't take many practice sessions for you to discover that not all landing approaches work out the way you'd like—even when you fly a good pattern, reduce power at the same place, maintain constant airspeed all the way down, and so forth. The culprit is usually the wind, and there's an easy way to take control and remedy the problem.

Let's assume for the moment that there is a 10-knot wind blowing straight down the runway, instead of the calm conditions that existed for our discussion of a "normal" approach and landing. As you take off and climb into this wind, you'll be moving across the ground at a lower speed (airspeed minus headwind velocity always equals ground speed), while the rate of climb remains the same. This means that the airplane will be at a higher altitude sooner (in terms of distance, not time), and when you turn onto the downwind leg, ground speed will increase, effectively shortening that portion of the traffic pattern. The first wind adjustment, therefore, should be to extend the takeoff leg a few extra seconds (experience will show you how much) so that there's enough time on the downwind leg to run through the GUMPS checklist and get set up for the approach.

The turn from downwind to base leg affords perhaps the best opportunity to adjust a power-off pattern for wind. The first time or two around the circuit may well be exercises in

Normal Takeoffs and Landings

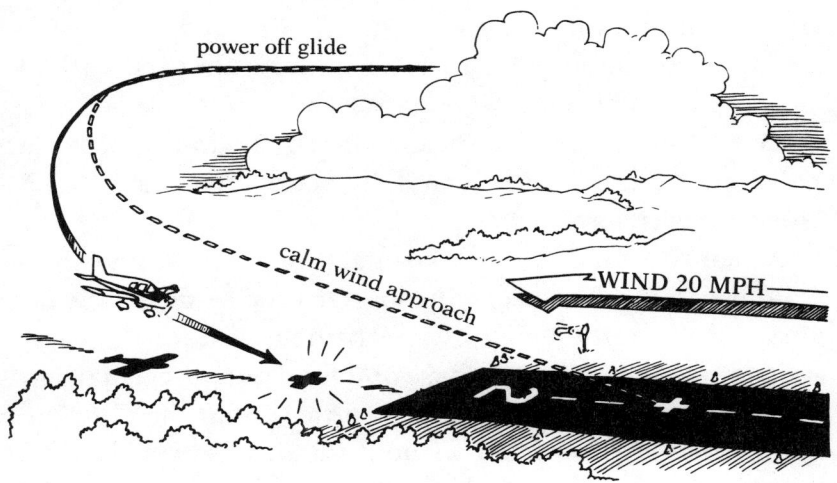

Every approach must be adjusted for the effect of the wind

cut-and-try until you develop a basis for judgment in this regard, but by all means, fly a consistent pattern. When everything else in the pattern is exactly the same (altitude, airspeed, power-down point, distance from the runway), you need only change the place at which you turn base. The steeper climb you encountered on the takeoff leg because of a headwind will act in a similar manner on final approach, causing the airplane to lose altitude at the same normal rate, but moving across the ground at a slower speed. In other words, if the airplane descends at the rate of 500 feet per minute, but that minute is not enough time to reach the runway at the existing ground speed, you're gonna come up short! There's no other way.

The problem can be solved by turning onto the base leg earlier than in a no-wind condition, thereby placing the airplane on the final approach leg closer to the runway. The glide path will be somewhat steeper, and there will be a strong temptation to fly at a higher airspeed, but resist—maintain the same pitch attitude, and the airspeed will take care of itself. This method of pattern adjustment is worthless if the pilot doesn't keep track of the downwind-to-base turn point; pick out a house, barn, lake, road, anything that provides a positive reference for that turn.

When the wind is much stronger, or when you wish (for whatever reason) to keep your pattern the same as it was in a no-wind condition, simply substitute thrust for the energy sapped by the wind. Turn base and final over the usual spots on the ground, then adjust the throttle so that the glide path looks just the way it did without wind. In effect, you'll be flying some small extra distance through the air (which is moving toward you) in order to get to the runway in the time allotted.

Those airplanes equipped with wing flaps have yet another option in this matter of adjusting the glide path to suit conditions. Wing flaps generally lower landing speeds (a good feature), but are also capable of steepening the glide path on final approach. With the added drag of the flaps, you'll have to point the nose down a bit more to maintain the proper airspeed, and that means a steeper approach—not a bad deal when your landing must be made over a row of tall trees at the end of the runway.

Experiment with patterns of various sizes, try out different combinations of airspeeds, wing flap settings, and power settings. Once you've mastered the normal power-off approach and landing, you should begin to explore other ways of getting the job done.

11

Abnormal Landings

WHY the term "abnormal"? Since the weight and flight characteristics of the airplanes expected to be used in Recreational Pilot training lend themselves so well to the power-off approach and landing, we've chosen that procedure as "normal." That leaves several other types of landings that can't be classified as high-performance or emergency landings, but which are sufficiently different from the power-off approach to rate separate attention. Hence this chapter, dealing with no-flap, engine-out, and wheel landings. (If you wonder why engine-out landings are discussed here, rather than in Emergency Procedures, ponder this: the only thing abnormal about an engine failure and a subsequent off-airport landing is the "runway"—*if* you have become proficient at normal, power-off approaches.)

THE NO-FLAP APPROACH AND LANDING

As explained earlier, wing flaps provide for a slower, steeper approach to landing. When there's a mechanical or electrical failure that renders the flap system inoperative, or when you want to do a "no-flapper" for practice, you are automatically committed to a flatter final approach profile.

Without flaps, fly a normal pattern up to the point at which you normally turn from downwind to base, but extend the downwind leg slightly (10 to 15 seconds) to accommodate the longer, flatter glide on final. Most light airplanes will perform admirably in a no-flap configuration at the airspeed used under normal conditions. If you feel more comfortable at a slightly higher airspeed, no problem—but remember that increases touchdown speed, which adds to the amount of runway required to stop the airplane.

Use power to adjust the glide path, and carry some power right through the flare and touchdown if it's needed. The pitch attitude that has served you well with flaps will probably also do the job when they are not available; you may want to experiment (under the watchful eye of your Flight Instructor, of course) to find the ideal attitude for touchdown. Without flaps, stall speed (and therefore touchdown speed) will be higher, meaning that the use of more runway to get stopped is inevitable. Plan ahead, and if the runway on which you'd like to land without flaps appears too short, go somewhere else.

ENGINE-OUT APPROACHES AND LANDINGS

When the powerplant on a single-engine airplane quits, there's no doubt what is going to happen: an unscheduled landing. An "emergency"? Yes, using the most literal interpretation, but it certainly doesn't mean an automatic disas-

Abnormal Landings

ter. If you've done your homework with regard to normal approaches and landings (here's another reason for practicing "normal" power-off approaches), an engine-out situation should present an emergency only as it regards the surface on which the airplane will come to rest.

There are a thousand variations on the engine-failure theme—altitude, airspeed, surface conditions, proximity to an airport, and so forth—which make it impossible to set down on paper the very best thing to do in any situation. But we can set out some general categories, which will provide a foundation for the application of common sense and good judgment at the time of an engine failure. These situations are:

- On takeoff, below 500 feet AGL.
- On takeoff, above 500 feet AGL.
- In cruise configuration, at pattern altitude.
- In cruise configuration, above pattern altitude.

Be aware that there are several cardinal rules which must be applied to *any* of these situations. First, and most important, FLY THE AIRPLANE! If you lose control, nothing else matters.

Second, establish the airplane in a pitch attitude that will provide its best glide speed; you are vitally interested in obtaining either the greatest possible gliding distance, or more time in the air for decision making ("best glide speed" is usually very close to "best rate of climb speed").

Third, troubleshoot *if you have time*. Try to figure out what has caused the engine stoppage—carburetor ice (very likely), fuel starvation (change tanks), whatever else is recommended by the airplane manufacturer.

Recreational Flying

Fourth, put the airplane down *under control*. This last item is just as important for your survival as the first; a good choice of landing site, a proper pattern, a good approach are all for naught if the airplane cartwheels or noses over or stalls out too high. Your airplane will do a creditable job of protecting you in a straight-ahead, level, low-speed encounter with the ground, but it's not built to take much of an impact on its side, or its nose, or upside down. Even if you have to land in the tops of the trees, *do it under control!*

ENGINE FAILURE ON TAKEOFF
(LESS THAN 500 FEET AGL)

In this situation, your best bet is to land straight ahead (or nearly so), since there won't be enough time to turn the airplane toward a better choice. *Attempting to turn 180 degrees and return to the field from which you have just taken off is almost always an invitation to disaster.* A great deal of altitude will be lost in the powerless turn, and in most circumstances, you will not be able to complete the turn before reaching the ground.

Light airplanes are straining in a normal climb: relatively low airspeed, full power, nose-high pitch attitude, everything working at maximum. When the source of thrust is suddenly removed, the transition from a climb to a glide must happen very quickly; there will be a significant loss of airspeed if you are slow in establishing the glide attitude, and the rate of descent will build rapidly.

Choose the landing area that appears best (more on that later), maintain best glide speed all the way to roundout (just as in a normal landing), and if your airplane is equipped

Abnormal Landings

with flaps, use them to lower the touchdown speed as much as possible.

Only very small turns (on the order of 20 degrees or so) should be considered when "straight ahead" is unacceptable; of course, it's wise to take a look at the terrain and obstacles off the end of the runway before you release the brakes for takeoff. An ounce of prevention . . .

ENGINE FAILURE ON TAKEOFF (MORE THAN 500 FEET AGL)

Your situation improves with every foot of altitude gained above the "magic number" of 500 feet (that's magic only because most light airplanes can't get turned around with less altitude; your airplane may require *considerably* more than 500 feet). Given the additional time to glide, you *may* be able to return to the airport, but you should experiment with your airplane and find out how much altitude is required. Keep in mind that if you took off into the wind (the normal procedure) you'll be landing with a tailwind, under stress, and things will appear totally different than on a normal upwind landing. Again, experiment; try some downwind landings (when the pattern is empty) so that when you need to do it, it won't be a total surprise.

If the terrain ahead is unacceptable, and you don't think you can safely return to the airport, consider yourself on a modified base leg for your engine-out "runway" when the engine quits. Depending on the prevailing wind direction, plan to turn left or right onto the final approach leg for a better field.

ENGINE FAILURE IN CRUISE CONFIGURATION (AT PATTERN ALTITUDE—APPROXIMATELY 1,000 FEET AGL)

Just like a good sailor, a good pilot always knows the general direction and strength of the wind. At only 1,000 feet above the ground with a dead engine, it's probably too late to figure it out; keep track of the signs as you fly. There are ample ways to determine wind direction, not the least of which is remembering what it was when you took off; this method obviously begins to fall apart as you fly farther from home. Other indicators are smoke from factories and fires, flags, wave or ripple patterns on bodies of water (even small farm ponds are helpful in this regard), and of course the old saw about cattle standing with their tails to the wind (you can believe this one or not, because sometimes they do, sometimes they don't!).

In any event, a general wind-direction determination is good enough to help you decide which way to land. As before, start the engine-out procedure by establishing the best glide speed, and while you're doing it, look out to both sides of the airplane. If there's a suitable field in the normal position (that is, halfway up the wing strut, or whatever visual reference you use for lateral positioning, and just about abeam), fly a normal approach. This is what you've been practicing all those patterns for!

If you're flying into the wind when the engine quits, turn 90 degrees to the left or right, and you'll be on a near-normal base leg. Should you be flying crosswind when the failure occurs, look to the left or right (depending on wind direction); you're already on base leg.

These suggestions will help you normalize your low-

Abnormal Landings

altitude engine-out patterns. Your chances of carrying off a successful forced landing will be greatly improved by flying the same kind of approach you've used in training.

ENGINE FAILURE IN CRUISE CONFIGURATION (ABOVE PATTERN ALTITUDE)

The extra altitude in this situation gives you an important advantage: instead of being forced (by lack of altitude or time) to choose a landing site very close by, there's time to expand your search and be more selective. Once again, be aware of the general wind direction (north, south, east, or west is good enough), but in this situation, you've plenty of time—at least, more than before—to pick out a suitable field within gliding distance and aligned with the wind.

The difference between this situation and the others lies in your ability to choose from a wider range of potential landing sites, and time to set up a normal pattern. Even if your choice is directly beneath the airplane, fly to the spot directly opposite the touchdown point, with the landing site halfway up the wing strut, airspeed and altitude right on the mark. From there on, it's a normal, power-off approach and landing.

Avoid the temptation to set up a long straight-in approach to a distant field; if it looks too far away, if you have a gut feeling that it's farther than you can glide, it probably is. Many pilots get themselves into engine-out trouble by trying to glide too far; if there's a good field close by, one you know you can make, don't pass it up.

For a cruise configuration procedure, then, pick a good field and set up a pattern to arrive on downwind opposite the intended landing spot at your normal pattern altitude, normal pattern airspeed, and a normal distance from the landing site. By doing as much of this as possible in a normal manner, you'll be able to call up those powers of judgment that have developed in the course of normal training.

LANDING AREA SELECTION

The regulations require that, in general, pilots fly no lower than would permit a safe landing in the event of engine failure. Sometimes, this rule is most difficult to observe, but it points up the importance of always having an emergency landing area in mind.

There are some general criteria; you should choose a landing site with respect to the wind (always land into the wind when possible, land crosswind or downwind only as a last resort), length of the field, type of surface (a smooth grass field is obviously best), and obstructions surrounding it (trees, power lines, buildings). Take time to bone up on farm-field use in your flying area, so you can tell the difference between a soft, just-plowed field (which just about guarantees a nose-over), and a near–runway quality strip of wheat stubble.

Roads and highways should be considered, but only as second choices to acceptable open fields. There's auto traffic to worry about, power lines and fences, and there might also be legal complications after you're down.

Power lines are a constant hazard for engineless aviators. If wires loom up in your path, try to fly under them if at all possible; remember that you are flying a glider, and if there's not enough energy present to get you over the wires (a very likely possibility), you may well stall out and drop into them anyway.

If there are no suitable fields available, land in the trees. Make it your very best full-stall landing, and try to touch down right on top of the trees, with the airplane under complete control.

Water landings are extremely hazardous at best, and you should not consider "ditching" unless there's nothing else to do. There is a wealth of information available on this subject, and if you are going to fly over or near water on a frequent basis, you'd do well to bone up on ditching procedures.

Finally, don't overlook *airports*. The next pilot who pulls off a successful engine-out landing in a farmer's field, only to find out that there is an airport right across the road, won't be the first. Know what's beneath as you fly along, and if the engine quits, *look around!*

AFTER THE LANDING

Assuming that you did everything right and got the airplane (and yourself) onto the ground in one piece, leave well enough alone—don't try to be a hero and fly the airplane out after it's been fixed. Swallow your pride and enlist the services of someone who really knows the airplane (preferably a veteran Flight Instructor); light airplanes are notorious for getting into fields they can't be flown out of. You've done a good job so far, so don't botch your record by attempting an operation beyond your skill level, or perhaps beyond the airplane's performance capabilities.

Recreational Flying

THE WHEEL LANDING

Unless your airplane is equipped with floats or skis, every landing will obviously be made on the wheels; but there's a special operation unique to taildraggers that has come to be known as "the wheel landing." Instead of the previously discussed normal (three-point) landing, a wheel landing consists of flying the airplane onto the runway at a higher airspeed, and touching down on the main wheels only.

Wheel landings are often used by veteran taildragger pilots as a means of coping with strong crosswinds (more airspeed equals more control authority), or sometimes just to exercise and show off their skill with the airplane. A wheel landing is a demanding technique that should not be attempted until you feel very comfortable in a taildragger, and after you have undergone a considerable amount of instruction in this operation.

The set-up for a wheel landing begins with a flatter final approach, occasioned by the slightly higher airspeed and partial power setting used throughout. As the touchdown point is approached, a near-level pitch attitude is maintained—no flare—and the rate of descent to the runway is controlled by slight changes in power. A lot of judgment (developed only by a lot of experience!) is required to anticipate when the main wheels will touch the surface; at that point, power is reduced completely, and a bit of forward pressure is applied to the control wheel or stick to keep the airplane from bouncing back into the air. As airspeed drops off, the tail will sink slowly to the runway, at which time it must be held there by full aft movement of the control wheel or stick.

You must remember that because a wheel landing is per-

Abnormal Landings

formed at a higher airspeed than normal, and because the tail is held off the runway for a relatively long time during which little or no braking can be used, considerably more runway will be used to get stopped. A wheel landing can indeed help a skillful pilot overcome the bad effects of a strong crosswind, but there's a price to be paid. A novice pilot should consider the wheel landing an advanced operation, one that must be properly taught and practiced until a safe level of skill is achieved.

12

Crosswind Takeoffs and Landings

IF YOU SHOULD EVER HAVE the pleasure of observing a blimp taking off or landing, you'll notice that the crew is extremely careful to operate directly into the wind (as if they had much to say about it anyway, since the blimp is the world's biggest weathervane!). The exceptionally large surface area and light weight of the blimp make it very susceptible to the force of the wind, even a mild breeze; unless it is aligned directly into the wind, it may well go kiting off downwind, literally out of control.

A Boeing 747—all 800,000-plus pounds of it—is little affected by a light wind on takeoff and landing, but you can bet that your training airplane—all 1,000-plus pounds of it—will respond, by comparison, as if it were the blimp. Fortunately, you have a lot more going for you in the way of controllability, and a surprisingly strong wind can be han-

Crosswind Takeoffs and Landings

dled when the proper technique and procedure are used. That's what this chapter is all about.

A well-executed crosswind takeoff or landing may well be the finest expression of the art of piloting an airplane; it requires more control inputs and constantly changing control pressures throughout. It's a lot like the skills demonstrated by a sailboat captain when he overcomes wind, current, and momentum to end a cruise with a gentle kiss of boat to dock.

DEFINITIONS FIRST

A crosswind takeoff or landing is one in which the wind, if not countered, will move the airplane sideways across the ground. Not only does this make it difficult to stay on the runway; a side load applied to the landing gear just before lift-off or immediately after touchdown may have disastrous consequences. With that in mind, let's expand the definition a bit; a *proper* crosswind takeoff or landing is one in which the pilot applies control pressures as required to complete the operation with no side loads imposed on the airplane.

The strength of a crosswind (the force trying to blow your airplane off the runway) is a combination of two factors: the velocity of the wind, and its direction. When the wind is blowing precisely down the center of the runway, it's all headwind or tailwind; but when the direction changes—even a few degrees—some amount of crosswind is present.

TAKING THE "CROSS" OUT OF CROSSWIND

Paint a picture in your mind: you're flying down the centerline of a long runway at normal cruise airspeed, five feet above the

Recreational Flying

pavement, in a flat, dead calm . . . absolutely no air moving. If you don't change heading at all, you would arrive at the other end of the runway directly over the centerline.

Now let's introduce a crosswind from the left, and try it again, with the objective of staying on the centerline. You wouldn't fly very far without noticing that the airplane is drifting to the right, and with no counteraction, you'd be considerably off course at the far end of the runway. One way to solve the problem is to turn the airplane to the left just enough to match the crosswind component; but if you allowed the wheels to touch the concrete in this condition (the airplane moving sideways with relation to the surface), there would be loud squeals of complaint from the tires at best, an upset of the whole machine at worst. There *must* be a better way of correcting for a crosswind, if you intend to land comfortably and safely.

With that in mind, imagine yourself having just turned onto the final approach leg, gliding toward the runway, with a crosswind gently pushing the airplane to the right. As soon

A typical crosswind approach and landing

Crosswind Takeoffs and Landings

as you detect the drift, bank to the left just enough to stop the sideways movement, and hold that bank angle. At the same time, press on the right rudder just enough to prevent yaw—keep the centerline of the airplane parallel to the centerline of the runway. When all of this is accomplished, you have redistributed the lift of the wings to counteract the crosswind, and have kept the airplane aligned with the runway for a no-side-load touchdown.

The same principle applies to a crosswind takeoff: the wings are banked in order to produce a horizontal component of lift equal to the force of the crosswind, and rudder pressure is applied as necessary to keep airplane and runway centerlines parallel.

So much for the theory. Here's how to put it to work.

CROSSWIND TAKEOFF PROCEDURE

There's one thing for certain—a pilot should never be surprised by a crosswind on takeoff. After all, you spent some time preflighting your airplane, noticed the effect of the wind while taxiing, and more than likely observed a wind sock or wind tee somewhere near the departure end of the runway. Forewarned is forearmed, so you can at least position the ailerons correctly before the takeoff roll begins. (Caution: since you'll be turning into the wind for takeoff, be aware of the increased potential for upset; the side force of the wind, plus the relatively high CG location and the centrifugal force of the turn, may add up to enough push to raise the upwind wing, and possibly turn the airplane over.)

Recreational Flying

Begin the takeoff roll just as you would for a no-crosswind operation, but be slow with the throttle, especially when the crosswind is from your left. In this situation, yaw to the left due to P force will be complemented by the weathervaning effect of the crosswind (also to the left), and a rapid application of throttle may produce more yaw than you are prepared to handle all at once! As airspeed increases, this problem will diminish as the rudder gains authority from the increased airflow around it. Conversely, a right crosswind may cancel out P force entirely; your job is to prevent yaw with the rudder, using whatever pressure is required.

As in a normal takeoff, establish the takeoff pitch attitude and allow the airplane to fly when it's ready; but be aware that in this attitude, the airplane will be rolling on only two wheels, and the weathervaning tendency is remarkably increased. You'll need to be alert, quick, and smooth with rudder inputs to maintain heading. (In stronger crosswinds, consider waiting a few seconds before establishing takeoff attitude, thereby allowing airspeed to increase and providing more control authority to counter the crosswind.)

After liftoff, the airplane will probably try to turn into the wind as a result of your aileron inputs; let it turn, and use that small heading change to help set up a crab angle to track straight down the runway. At this point your best indicator of drift correction is the runway itself, so look out and down to see how you're doing. Establish a heading that will counteract any drift, get coordinated (wings level, no yaw), and continue climbing.

During the initial part of the takeoff roll, it's difficult to determine how much the crosswind is pushing sideways, since the airplane is pretty firmly anchored to the runway by the friction of tires against pavement. You can solve this problem by using whatever aileron displacement is neces-

Crosswind Takeoffs and Landings

sary to keep the wings level throughout the takeoff roll. When operating in a strong crosswind, it's a good idea to start the takeoff with full aileron into the wind, then take out whatever you don't need as airspeed increases and control authority builds. Use just enough aileron to keep the wings level.

CROSSWIND LANDING PROCEDURE

A well-executed crosswind landing begins well back in the approach phase; perhaps even as far back as the preflight weather briefing, when wind direction and velocity at the destination airport can be at least roughly determined. (The "preflight briefing" can be anything from a formal visit to the FSS to a wet finger held up in the breeze, and "destination airport" may well be the field from which you depart.) If nothing else, the amount of crosswind you'll have to contend with on landing should become apparent as you fly the traffic pattern, and you should be getting yourself prepared mentally as soon as a crosswind is detected.

To begin a crosswind landing, fly a normal pattern, and when lined up on final, stop any drift by banking into the wind, all the while maintaining runway alignment with rudder pressure. This is indeed a crossed-control situation (wheel or stick left, rudder right, or vice versa), but for good crosswind correction, it is not abnormal. You must override the strong desire to return the wheel or stick to neutral; this tendency is no doubt the result of being taught that the wings must be level for touchdown. That's true, but only for normal, no-crosswind landings. Establish this in your mind at the outset: proper crosswind landing technique will result in the airplane touching down on one main wheel, with the wings banked into the wind.

As you proceed down the final approach course, wind direction and velocity will probably change, requiring your constant attention, and most likely, constantly changing control inputs and pressures. Remember, the drill is to use enough aileron pressure to stop the drift and enough rudder pressure to keep the airplane aligned with the runway.

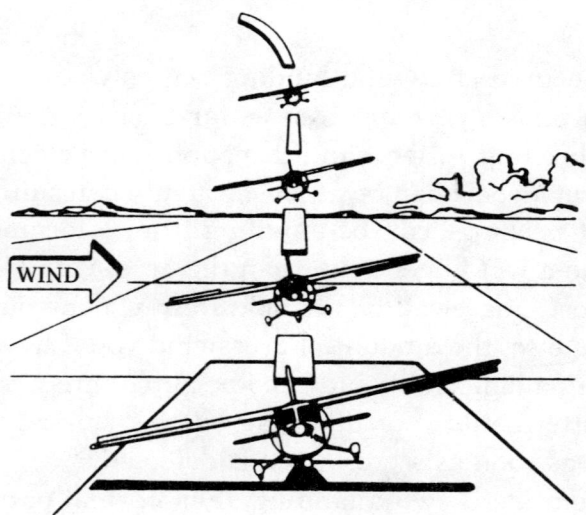

A proper crosswind landing takes place on the upwind wheel first

The task gets more demanding in the roundout and touchdown, since diminished airspeed means less control authority. You must continue to stop drift with the ailerons and maintain runway alignment with rudder pressure, and *full displacement* of the controls may become necessary. Here's

Crosswind Takeoffs and Landings

where "Taylor's Rule" applies: Do whatever you need to do (full movement) with whatever's available (flight controls) to make the airplane do what you want it to do (stop the drift and keep it straight).

It should be clear by now that if proper crosswind technique is used, the wings will be banked into the wind at landing; therefore, to assure that no side load is present at touchdown, you will actually land on one main wheel, the one on the upwind side of the airplane. As airspeed drops, increase aileron pressure to prevent drift until there's no more control travel; when aileron authority is completely gone, the wings will drop gently to a level, both-wheels-on-the-ground attitude, and you should then revert to the proper crosswind taxiing technique.

Turning off the runway may present the same potential upset problem you encountered when getting ready for takeoff, and the same precautions and procedures apply. On a day when the wind is strong and gusty as well, you must redouble your alertness and dedication to good technique.

The most persistent crosswind problem encountered by beginning pilots (especially in very lightweight airplanes) is failure to use the flight controls to the extent required. Resolve at the outset to use as much control displacement as you need, and don't worry about breaking anything; the control cables are stronger than you are!

Failure (or inability) to detect drift is another major student-pilot problem. If you have trouble visualizing sideways motion of the airplane, pick a crosswindy day and fly (with your instructor along) the length of the runway just off the ground with no correction, so you can see what drift looks like. Then come back for a second run, this time lowering the wing into the wind and maintaining alignment with the rudder, so you can see what drift *correction* looks like.

CROSSWIND LIMITS AND EXTREME CONDITIONS

Every airplane has a crosswind limit; that is, a point at which the pilot is doing everything possible to overcome drift (full aileron, full rudder), but the airplane continues to drift off the runway. When this limit is published in the Pilot's Operating Handbook or Owner's Manual, treat it as an advisory only. Stay well below the quoted value; it's usually a number developed by a test pilot, who is well paid to find out how far he can push the airplane.

Your personal crosswind limit should be very conservative at first, and should grow with your aeronautical experience, confidence, and ability. In any event, if you find that full (or nearly full) control displacement is required on final approach and the airplane is still drifting, by all means go around and think of something else. If full control is required at approach speed, there's no way you'll be able to hold the airplane on centerline at the lower speed required for landing.

The obvious solution to this problem is to fly faster, thereby providing more control authority to overcome drift. An admirable suggestion, but you've got to slow down to land sooner or later... then what? There are advanced crosswind techniques, ways to handle this problem, but they are beyond the level of this book. When in doubt, go around, then find another runway or another airport; better yet, be

Crosswind Takeoffs and Landings

aware of the weather situation, and don't be caught aloft in the first place.

Excessive airspeed during a crosswind landing (or any landing, for that matter) sets you up for either one or both of two problems: wheelbarrowing or a ground loop. In the first case, limited to tricycle-gear airplanes, additional speed causes the wings to develop enough lift to take the weight off the main wheels; steering becomes next to impossible with only the nosewheel in contact with the runway, and a rapid excursion off into the grass (or worse) is often the result. There's only one solution—SLOW DOWN. And if the slower speed means that you can't control the drift, go somewhere else to land.

A ground loop, in which the airplane tries to swap ends during the landing roll, is also a product of excessive airspeed, plus less-than-adept handling of the rudder. A ground loop may be performed by pilots of either tricycle or tailwheel airplanes, but it's much more likely in a taildragger. With the CG located considerably farther aft of the main wheels (the pivot point when on the ground), a taildragger tends to continue turning once it starts... and if there's enough extra momentum (speed) involved, the pilot may find himself just going along for the ride. Ground loops are usually more embarrassing than dangerous, but they can be prevented by controlling speed and by using proper rudder technique.

Suppose that you are faced with a situation in which the crosswind component is excessive, but you must land. Increase the approach speed slightly to improve control authority, then experiment to find the minimum airspeed for satisfactory control. In a taildragger, consider a wheel landing; but remember that you'll be riding on the main wheels for a longer period of time, vulnerable to all the effects of weathervaning. Either of these techniques will increase the distance required for landing.

SUMMING IT UP

When taking off or landing in a crosswind, use aileron to stop or control drift, use rudder pressure to maintain your alignment with the runway. Expect to land on the upwind wheel, and don't be bashful with the flight controls. Do whatever you need to do with whatever's available to make the airplane do what you want it to do.

13

Airport Operations

THE UNCONTROLLED AIRPORT (one with no control tower in operation) is truly the domain of the Recreational Pilot. Unfortunately, it is also the domain of most of the light-airplane midair collisions, and most of these mishaps can be traced to lack of or improper communications, or to pilots who don't observe the flight rules that apply to uncontrolled airport operations. Since the Recreational Pilot is not likely to be flying an airplane capable of radio communications, the latter is most important, and this chapter will deal with the rules, regulations, and procedures for flight operations on and in the vicinity of an uncontrolled airport.

Keep in mind that since air traffic control at an uncontrolled field is rather informal, and completely in the hands of the pilots involved, it's more important than ever to know and follow the rules. The keystone of safe self-controlled operations is simple: if each pilot adheres to the rules, knows where to look for other airplanes, and maintains vigilance so as to see and avoid other aircraft, there can be no midair collisions ever again. What an admirable objective for all of us involved in the operation of airplanes!

Recreational Flying

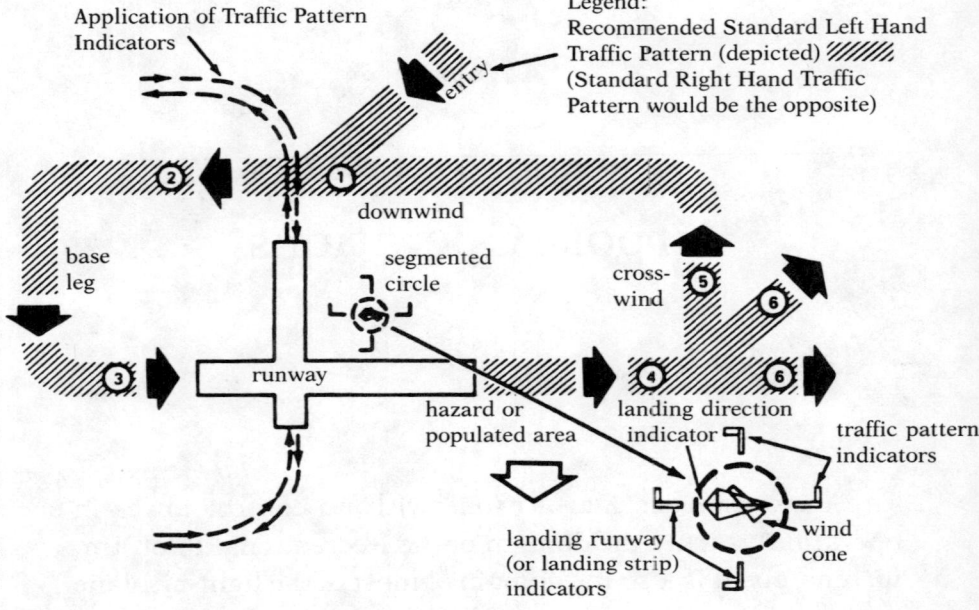

① Enter pattern in level flight, abeam the midpoint of the runway, at pattern altitude. (1000' AGL is recommended pattern altitude unless established otherwise.)

② Maintain pattern altitude until abeam approach end of the landing runway, on downwind leg.

③ Complete turn to final at least ¼ mile from the runway.

④ Continue straight ahead until beyond departure end of runway.

⑤ If remaining in the traffic pattern, commence turn to crosswind leg beyond the departure end of the runway, within 300 feet of pattern altitude.

⑥ If departing the traffic pattern, continue straight out, or exit with a 45° left turn beyond the departure end of the runway, after reaching pattern altitude.

Recommended (strongly!) airport traffic pattern entry and exit procedures

Airport Operations

RUNWAY SELECTION

The most important factor in choosing a runway for takeoff or landing is wind direction and velocity, in the interest of optimum airplane performance. Almost all airports will have some sort of wind direction indicator—a wind sock or wind tee—and many operators have installed electrical devices that sample the wind from atop a small tower, or perhaps the top of a hangar, and provide a readout of direction and speed on dials inside the operations building. If nothing else, you can always hold up a wet finger, or observe the behavior of smoke or flags in the vicinity of the airport. When the wind is nearly calm, you'll see veteran aviators harvest a handful of dry grass or dust, toss it into the air, and thereby determine whence the wind bloweth.

From the air, inbound to the airport, you have the advantage of wider horizons, and therefore the ability to sample more wind indicators. The wind sock or tee on the airport is still probably the most reliable, but in its absence (or when you wish to plan farther ahead, from a point too far away to see the airport), you should be able to find some smoke, flags, ripples on a lake or pond—there's always something affected visually by the wind.

Your fellow pilots provide yet another source of runway direction information. Especially when there are a number of airplanes in the traffic pattern playing follow-the-leader, it's likely that the first pilot into the air made the proper choice, and you should go with the flow. This procedure must allow for others' errors, of course, and if it is apparent to you that everybody is on a merry-go-round in the wrong direction, don't attempt to reverse the circuit by flying an opposite-direction pattern. Wait until the pattern empties or

Recreational Flying

go somewhere else, but don't fly in the face of other airplanes—it's hardly worth the risk.

A calm or near-calm wind condition presents the most hazardous runway-selection situation. With the active runway up for grabs, you must assume that the airport is operating in accordance with Murphy's Law of Runway Selection: if it is possible for someone to choose the wrong runway, someone *will*. Exercise *extreme* caution, and have a Plan B ready if you spot another airplane flying the opposite direction. Don't dispute the right-of-way.

OUTBOUND PROCEDURES

Once you have decided which runway to use for takeoff, give some thought to the safest way to get to the departure end. Never taxi on an active runway if you don't need to; why increase your exposure? When a taxiway is available, use it. If there's no taxiway, but solid, smooth grass alongside the runway, use that for taxiing.

In those cases where the runway must be used to get to the other end, minimize the risk by taxiing as close to the edge of the pavement as possible, and resolve to always turn your airplane through a full circle to provide a complete scan of the pattern before moving onto the runway proper (airplanes with full-view canopies excepted).

Some airports (notably the smaller ones, where pavement is at a premium) have no provision for a runup pad at the end of the runway, so complete your before-takeoff checks prior to taxiing onto the active runway. By doing this, you'll be ready to go when you reach the end, and again lower your exposure to a potential collision.

Airport Operations

The traditional (and regulatory) standard for traffic pattern direction at uncontrolled airports calls for all turns to the left, but due to noise considerations, or obstructions, or whatever, certain airports use nonstandard patterns. You are required to be aware of this situation, and when your plan is to remain in closed traffic, observe whatever type of pattern flow is indicated.

There are frequent occurrences of nonstandard departure procedures (such as "climb to 2,000 feet before turning," or "turn to heading 140 as soon as practicable after takeoff," for example), and once published, such procedures must be observed. You can bet your bottom dollar that the pilot who gets involved in an incident or accident because he didn't observe the nonstandard procedures will have to answer for his actions—sometimes very expensively.

When remaining in the pattern for takeoff or landing practice, use the crosswind leg to visually clear the entry and downwind legs. Here's where pilot cooperation really pays off, for if everyone who enters a traffic pattern does so properly, and if pilots remaining in the pattern know where to look for inbound airplanes, the risk of midair collisions is significantly reduced.

For the same reasons, use the base leg to clear the final approach course. Statistics show that the area from mid-downwind to touchdown is the most hazardous with respect to midair collisions. You can do a lot to prevent this problem by making sure that no one else is lined up for landing.

The safest way to exit a traffic pattern is, in any case, to remove yourself from possible conflict with other airplanes as soon as you can. Sometimes (in the absence of specific departure procedures for the airport), flying straight out from the runway does the job best; or you may prefer to turn to the right or left at a safe altitude. In any event, fly away

Recreational Flying

from the established traffic pattern as soon as it's safe and practical to do so. While the "straight-out" or "turn-away-from-the-pattern" departures are those officially suggested, there are no firm rules in this regard; you're very much on your own.

INBOUND PROCEDURES

After a few ground-training sessions, and a few hours in the air, you'll have a good working knowledge of the traffic pattern practices and regulations at your home airport. But when you begin to use your airplane for more than just local training flights, it becomes absolutely necessary to know what to do at a strange airfield, especially in light of other pilots' dependence on your doing things the proper way.

Determine the pattern layout as soon as possible when inbound to an uncontrolled airport. With no radio, this usually means flying overhead—at least 500 feet above traffic pattern altitude—and checking for other airplanes as well as for the standard traffic pattern indicators on the ground.

With the landing runway decided upon, visualize the downwind leg and plan the pattern entry. In the interest of safety (remember, you want to be where other pilots *expect* you to be), the entry leg could be the most important of all, because it provides the opportunity to scan the entire pattern for other airplanes. Give yourself plenty of room for the entry leg, and plan to enter the downward leg no later than the mid-point of the runway; entering any farther along makes the downwind too short to get your before-landing chores accomplished. If you have to fly an extra minute or two to position yourself properly for the entry, so be it.

It's also very important to be at traffic pattern altitude on the entry leg, to preclude the possibility of descending onto another airplane in the pattern—very few of our flying machines afford good visibility straight down.

ODDS AND ENDS OF AIRPORT OPERATIONS

The right-of-way rules that govern all of aviation come into particular significance in the vicinity of an airport, when airplanes are necessarily operating in close proximity to each other, and when the flow of traffic is not being monitored by a control tower. In general, when you are flying a faster airplane, the slower fellow has the right of way, and you must pass well clear. If you spot another airplane on a converging course, the airplane on the right has right-of-way, but don't depend on it, and don't hesitate; as soon as you detect a potential conflict, turn, climb, or descend—do something to *defuse* the situation.

With specific reference to airport operations, when two airplanes are approaching the same runway for landing, the lower one has the right-of-way; but this rule doesn't give the lower pilot the right to cut in front of the other. There's a Golden Rule with regard to right-of-way: when you see a potential traffic conflict shaping up, move yourself out of the way *right now*, and argue (if you must) about the legalities of the situation when you're both safely on the ground.

Very few uncontrolled airports have any kind of spectator control. People, dogs, kids, and cars are generally permitted the run of the ramp, and there are numerous hazards inherent when such as these are mixed with airplane movements. Be especially aware of the invisibility of rotating propeller blades, and the masking of airplane engine sounds by other noises in the airport.

Finally, whenever you are flying in the vicinity of an uncontrolled airport, remember that pilots of radio-equipped airplanes rely very heavily on voice communications and broadcasts of others' positions and intentions. The nonradio folks have just as much right as anyone to use that airspace, but without radio, you've got to double and redouble your awareness of other airplanes on and around the airport. If, after your first few flights, you don't have a sore neck from looking around constantly, you're not doing the job right.

14

Airspace Considerations

ONE of the primary responsibilities of the Federal Aviation Administration is the management of the airspace, so that everyone who is qualified has adequate access to this vast national resource. The primary concern is the prevention of midair collisions, and one way to achieve this objective is to set aside certain segments of the airspace for users in accordance with their qualifications and needs.

For example, pilots with instrument ratings need the assurance that they are safely separated from other airplanes when flying inside the clouds. So non-instrument pilots are prohibited from operating in weather conditions that are less than the basic VFR minimums, and in certain types of airspace no matter what the weather.

The Recreational Pilot, not trained or equipped for radio communication with the Air Traffic Control System, is even more limited. Fortunately, the airspace you may use is much greater than that from which you are restricted, so the easiest way to stay legal is to develop a good understanding of those areas you must avoid. In other words, the Recreational

Recreational Flying

Pilot may fly anywhere he pleases, except.... You'll do well to have a Sectional Aeronautical Chart and its comprehensive legend handy from here on, to identify and become familiar with the markings and dimensions of airspace segments that are off-limits for you.

GENERAL RESTRICTIONS

The Recreational Pilot may not fly in controlled airspace when the weather conditions are less than basic VFR minimums (see Chapter 1). In addition to the obvious concern for the instrument pilots who use this airspace when the weather is bad, there's a very high probability of an untrained pilot getting into serious airplane-control trouble when the visibility drops.

For physiological reasons (lack of sufficient oxygen pressure), a general altitude limit is also imposed; you may not fly higher than 10,000 feet MSL, or 2,000 feet above the surface, whichever is higher. This means that you could cross directly above the peak of Mt. Whitney at an altitude of 16,495 feet and be completely legal ... but not necessarily smart. Each one of us human beings, knowledge and training notwithstanding, becomes adversely affected by lack of oxygen above 10,000 feet; if you must find out what it's like, do so in a controlled environment such as the physiological training courses offered by the FAA and certain military installations (contact the local Flight Standards District Office of the FAA for this information).

Airspace Considerations

SPECIFIC RESTRICTIONS

Controlled Airports

A controlled airport is defined as any airport at which a control tower is in operation, and of course that means that radio communication between controller and pilot is necessary to accomplish the objective. Without the training and equipment to do this, the Recreational Pilot may not operate at a controlled airport. In many cases, especially at the less-busy terminals, the control tower is a part-time operation, and the airport then becomes available to Recreational Pilots; but bear in mind that except for very early on summer mornings or very late on summer evenings, the nontower hours will also likely be hours of darkness. You should probably exclude all controlled airports from your list of available facilities.

The controlled-airport restriction applies to more than takeoffs and landings. In order to do their work properly, controllers need to have some assurance that they are talking to all of the aircraft within a given space. With this in mind, the *Airport Traffic Area* was created, and consists of a cylinder of airspace 3,000 feet tall (starting from the surface) and extending for five miles in all directions from the center of the airport. In addition, airports which have any kind of Air Traffic Control facility and an instrument approach procedure will be further endowed with a *Control Zone*, also five miles in radius in most cases. So, the Recreational Pilot can stay legal and enhance safety by avoiding *all* controlled airports by at least five miles.

Recreational Flying

At certain terminals (the very busy ones), controllers are unable to separate traffic safely unless they have both radio and radar contact with the aircraft flying to and from the airport. In these situations, a *Terminal Control Area* (TCA) is established, usually five miles in radius at the surface and stepped upward and outward to perhaps 8,000 feet and 25 to 30 miles; less busy airports sometimes feature a *Terminal Radar Service Area* (TRSA) or *Airport Radar Service Area* (ARSA). These areas were created because of the increased number of airplanes operating therein and the attendant increased hazard; TCAs, TRSAs, and ARSAs need to be included in the "off-limits" airspace for the Recreational Pilot.

SPECIAL-USE AIRSPACE

We've been using the term "restricted" rather loosely until now, referring to those portions of the airspace in which Recreational Pilots may not fly because of communications or weather limitations. But there are several additional types of airspace which are more specifically defined and limited; these are the so-called Special-Use areas, and the restrictions vary in accordance with the activities they contain.

Prohibited Area

Most restrictive of all, a Prohibited Area means exactly what it says: no one is permitted to operate aircraft in that airspace. Usually established with national security in mind, Prohibited Areas are found in several places around Wash-

Airspace Considerations

ington, D.C., and other very sensitive locations. Don't even *think* about flying close to a Prohibited Area!

Restricted Area

You'd readily admit that attempting to share the airspace with such flying objects as artillery shells, rockets, guided missiles, cannon-firing fighter aircraft and the like could be hazardous to your health. That's the primary reason for the establishment of Restricted Areas. They are easily identified on the charts, are always numbered (R-1234, for example), and are sometimes available for use by airplanes just "passing through." You have the option of calling the nearest Flight Service Station to find out if a particular Restricted Area is in use, in which case you should avoid that area altogether.

Many Restricted Areas have floors above the surface, which permit light airplanes to operate safely underneath, but most of them have very high ceilings. In any case, a listing on the appropriate aeronautical chart will indicate the dimensions and use times of each Restricted Area.

Military Operations Area (MOA)

From the Recreational Pilot's point of view, an MOA is just another Restricted Area, but in this case the potentially hazardous activity consists of "Uncle Sam's finest" at work. There's no law against flying through an MOA—even while it's in use—but all VFR pilots are well advised to contact the nearest Flight Service Station for current information, and to use *extreme* caution and vigilance. Dimensions and times of use are stated in the appropriate listing on aeronautical charts.

Alert Area

Here and there on the charts (mostly in the southern states, where the flying weather is consistently good), you'll notice an Alert Area, defining a section of airspace in which a high volume of flight training is being conducted. The potential hazard of many airplanes operating in a confined area means that all the pilots involved need to exercise more caution than usual.

Special Conservation Area

Because the noise of airplanes (and sometimes their very presence) is disturbing to certain types of wildlife, federal and state authorities have identified nesting grounds and specific habitats for protection. In these cases, you'll notice a distinctive marking on the charts, accompanied by a notation describing the restrictions (usually 2,000 feet or so above the surface) applicable to all aircraft.

Temporary Flight Restriction

Natural disasters, major accidents, and other significant events tend to attract aerial sightseers—after all, an airplane provides a rather unique view. However, with the knowledge that some aircraft operations are essential therein, the FAA has reserved the right to place such an area off-limits to all other pilots. There are teeth in this one; when Temporary Flight Restrictions are placed in effect, that area comes under the regulation of FAR 91.91.

Airspace Considerations

A Temporary Flight Restriction will normally indicate the airspace 2,000 feet above the surface within five miles of the site; the exact dimensions will be found in NOTAMs (NOtices To AirMen), available at a nearby Flight Service Station. Permissible flight activities are usually commercial in nature, which effectively precludes Recreational Pilots.

15

Basic Aerial Navigation

THE ANNUAL MIGRATION of most birds to some other part of the country may be mostly instinctive, but did you ever stop to think that perhaps they just want to fly somewhere different for a change? You may find yourself in the same situation; bored by the traffic pattern and the local flying area, feeling the need to fly away, to find out what's over the horizon.

Even the shortest trips require some navigational sense, and some training. A lost bird can always land in the nearest tree, refuel with worms or seeds or bugs, and continue on its way; but a human aviator's "nearest tree" must be an airport. The embarrassment notwithstanding (no pilot likes to admit that he got lost), safety demands that you know where you are, where you've been, and where you're going.

That's what this chapter is all about, but on a very basic level. We'll deal with principles only, mostly because there are so many acceptable techniques and procedures, and those should be taught by the person who will have most to do with the application of them, your Flight Instructor. In

Basic Aerial Navigation

any event, if you understand the principles of aerial navigation, you'll be able to make sense of whatever tools are available, and more importantly, you'll be able to navigate safely from the proverbial Point A to Point B and back again.

AERONAUTICAL CHARTS

The cornerstone of nearly all types of navigation (aerial or otherwise) is a good chart. The government has developed a great deal of expertise in producing aeronautical charts for every type of flying. For the Recreational Pilot, the Sectional Chart is best for everyday use, with a scale of 1:500,000 and plenty of recognizable detail. There are currently thirty-seven such charts that cover the conterminous United States, and additional charts for Alaska and Hawaii.

Because the colors on these charts are very meaningful, and because you should have a current chart for your own local area, it would be a good idea for you to put a bookmark right here, go to the airport, and purchase a sectional. Besides, you can casually spread it out on the coffee table, where it makes a great conversation starter, as in, "Oh, do you fly?"

You'll notice that terrain elevation is indicated generally by means of coloration; green is sea level, and the higher the ground gets, the browner it becomes. Contour lines are generally drawn in 500-foot increments, and major peaks and point elevations are noted in feet above sea level. Each latitude-longitude square (marked by the vertical and horizontal black lines) contains a bold indication of the highest obstruction in that block; for example, "47" means that somewhere in that square is a mountain, tower, ridgeline, or building that tops out at 4,700 feet MSL.

Chart features will leap off the paper after you know what to look for; airports, highways, lakes, cities, and towns are well documented in the legend, always found on the front panel of every chart. There's no need to memorize any of this, since the most-used features will become old friends in a short while, and with the chart always handy in the airplane (never leave the home airport without a chart!), you've a quick reference for the features that are unfamiliar.

AERIAL NAVIGATION FOR THE RECREATIONAL PILOT

Pilotage

This is the most fundamental of all the ways to get an airplane from one place to another. It was the only way for pilots to navigate before they had charts, and it's still fashionable (and proper, and safe) to use pilotage in those situations where you know the route like the back of your hand. Simply put, pilotage is the technique of flying from one landmark to the next using nothing but visual references.

If, for example, you wish to fly from your home base to the airport at the next town to the east ("Eastville"), and you know that a large lake lies in between, take off, turn toward the sunrise, and climb until you can see the lake. Over the lake, look for the town and fly to it, find the airport, and land—mission accomplished. Your knowledge of a highway that connects the two towns helps even more; simply follow the road. (That, by the way, is the real meaning of IFR—"I Follow Roads.")

Pilotage is so fundamental that almost no planning is necessary (assuming that your trip is a short one, and you won't be concerned about adequate fuel supply), and you can safely head out using a rough direction—south, east, northwest, and so forth. But for practice, develop the habit of referring to a chart on even the shortest journeys, and make it easy for yourself; fold the chart to a convenient size (never cut a chart) and turn it so that you're looking at chart features with a real-world orientation (a good pilot learns quickly to read upside down!).

In the finest tradition of pilotage, you should literally fly from one landmark to the next, and it's vital for you to keep track of those landmarks as you go. Suppose that your first landmark is that lake between the two towns in our earlier example; when over the real lake, put your finger on the chart lake and keep it there until you can see the next town. So, in addition to flying from one landmark to the next, you should fly from one fingerprint to the next; that way, you'll always know where you have been.

Dead Reckoning

Rather than attempt to change a firmly established error in aviation's vocabulary, we'll go along with the word "dead." But you might like to know that it's a corruption of "deduced," and refers to the process of computing your airplane's future position on the basis of current knowledge. With that bit of verbal housekeeping out of the way, let's navigate.

Recreational Flying

In its simplest form, dead reckoning ("DR" from now on) might consist of a pilot knowing (1) that the air between his home base and Eastville is perfectly calm, (2) that his airplane travels through the air at 100 miles per hour, (3) that the direction from his home base to Eastville is exactly 90 degrees, and (4) that the distance is exactly 50 miles. With this knowledge (and a little calculation), the pilot deduces that if he flies eastward for thirty minutes, he'll arrive over Eastville.

It's not all that simple, of course. "Heading toward the sunrise" is hardly the kind of accuracy you need to be sure you're going in the right direction. A navigational plotter is used to determine, to the nearest degree, the direction of the course line you draw on the chart from departure to destination. The vertical black lines used to plot direction are referenced to the North Pole, but your airplane compass senses direction with relation to the *magnetic* north pole, which is considerably displaced. The difference is known as "variation," and you'll learn during ground training just how to find and apply the difference.

The other major factor in DR is the wind, and its effect on the flight of an airplane. Even when there are no breezes stirring on the ground, you can bet that there is *some* wind aloft. On some trips, the wind will be blowing from directly behind (a tailwind) or from directly in front (headwind), and in these situations, only your ground speed will be affected. With a tailwind, you'll move over the ground at your airspeed (100 MPH in our example) *plus* the speed of the wind; with a headwind, groundspeed will always be your airspeed *minus* the velocity of the wind.

Things get even more complicated when the wind blows from one side or the other, and this is usually the case. Now, in addition to the headwind or tailwind effect, the wind will

Basic Aerial Navigation

attempt to move the airplane somewhat sideways across the ground; if you are to make good your intended course, the airplane must be turned into the wind just enough to offset the sidewind, or "crosswind" as it's known in aviation. (You're accomplishing, en route, the very same thing we discussed in Chapter 12, "Crosswind Takeoffs and Landings.")

There are six factors in the dead reckoning navigation problem: airspeed, groundspeed, wind direction, wind velocity, course over the ground, and heading (the direction your airplane is pointed). In a flight-planning situation, you'll have four of these available: *course* has been plotted from the line on the chart, *airspeed* is determined either from past experience or reference to the airplane handbook, and *wind direction* and *velocity* are obtained in the form of a forecast from the Flight Service Station. By using a navigational computer, you can find out how much you should turn the airplane to counter the crosswind effect, and what the resultant groundspeed will be.

Now, armed with all that planning information, you are ready to figure out how long you should fly on a given heading in order to reach Eastville. In a totally pure environment (one in which everything worked out precisely as planned), you could leave your home base, fly for the prescribed number of minutes on the computed heading, look down, and expect to see Eastville Airport directly below. Don't count on it, however; winds almost always vary from the forecast (not to blame the forecasters; meteorology is a very inexact science), and you'll need to adjust as necessary along the way.

Your instructors (both ground and flight) will spend considerable time explaining and teaching the detailed procedures and computations necessary for good, reliable DR navigation.

Recreational Flying

Orientation Procedures

Oh, it will happen to you; you can count on it. There are two kinds of airplane pilots—those who have been lost, and those who will be!

As a matter of fact, your Flight Instructor will probably allow you to become "disoriented" early in your cross-country training, just so you'll know what it's like, and how to remedy the situation. There are so many reasons for disorientation, and so many ways to get unlost, we can only discuss the principles here; your instructor will cover specifics in considerable detail.

Good preflight preparation earns its keep when a pilot gets lost (let's call a spade a spade). Winds aloft don't change all that much over a short distance, so if you know what time your next checkpoint should appear, hold your heading religiously, fly out the time, and *look*—up, down, out to both sides, including directly under the airplane.

If there's still nothing recognizable, turn 90 degrees and fly a short arc to the left, then back to the right. In most such situations, you'll be within a reasonable distance of your checkpoint, and overflying a small area in the vicinity will usually turn up something worthwhile. Look for prominent landmarks such as large lakes with conspicuous shapes, major highways, multiple-track railroads, and so on. You'll often find the names of small towns and villages painted very prominently on water tanks, rooftops, factory signs, and the like. You may have to descend to minimum altitude to read the signs, but if it reorients you, that's the name of the game. Pick up the pieces and continue on your way.

If your "disorientation" proceeds to actually being lost, look for an uncontrolled airport, land, and find out where you are. Chances are that as you taxi to the ramp, you'll spot

Basic Aerial Navigation

a sign that will name the town, and save you the embarrassment of asking.

Completely lost, and with no airport in sight, your best bet is to follow a highway, railroad, or river (preferably downstream) until you come to a town, most of which have an airport of some sort on the outskirts. As soon as you spot an airport that appears safe in terms of runway conditions, length, and crosswind, *land—and find out where you are. 'Tis no sin to confess.*

16

Meteorology for the Recreational Pilot

ALL AVIATORS are concerned about the weather, because we're involved in it, good or bad, whenever we leave the surface of the earth. A sunny day with light winds and unlimited visibility presents little in the way of obstacles to flight; but when there are low clouds, restricted visibility, or high winds, pilots—and especially Recreational Pilots—need to reassess their need to fly that day.

The Recreational Pilot's weather limitations are perhaps more important than any others, for several good reasons: first, you will be flying an airplane that is not intended to joust with meteorological windmills; second, your training will not include an in-depth study of weather since you'll be operating very close to home base; and third, there is no requirement for you to receive any training in the techniques and procedures for getting yourself out of a no-visibility situation—that is, an encounter with instrument conditions. Rest assured that should you elect to continue your aviation education toward a Private Pilot certificate, you will spend a lot more time studying weather. For now, however, the basics will suffice.

CLOUDS

They may be called cumulus or stratus or cirrus, but clouds—no matter what their type—are nothing more than moisture vapor that has become visible because of condensation, usually the result of lowered temperature. All air contains some moisture, visible or not. One of the measures of moisture is *relative humidity*—the amount of water vapor in the air, expressed as a percentage of the moisture that air will hold as invisible vapor. We're interested in the air's water content in relationship to its temperature, because condensation will take place in any parcel of air when the temperature reaches a certain value.

As the temperature drops, the ability of the air to hold water as invisible vapor decreases. When the air is very moist (high relative humidity), the temperature needs to be lowered only a small amount for condensation to occur, and clouds form readily. Very dry air must be cooled a great deal before any of the water vapor becomes apparent to the eye. But regardless of the amount of moisture, the temperature at which condensation takes place is known as the *dew point*, and information about it is furnished as part of all hourly aviation weather observations. You can hang your hat on this fact; when the air temperature reaches its dew point, some sort of condensation takes place, and the moisture becomes visible.

Any process that causes the temperature to drop to the dew point can be responsible for cloud formation: radiation cooling after sunset, when the air close to the earth is cooled by contact with the cold ground, or perhaps an air mass being lifted over a mountain range and cooling as it rises and expands. In any case where the air is cooled to its dew point, condensation occurs. By the same token, when the temperature somehow gets back above the dew point, the clouds begin to disappear, which explains why early-morning fog goes away as the sun warms the air.

There are two cloud forms of primary interest to the Recreational Pilot—cumulus and stratus. Cumulus cloud formations are heaped-up masses, cauliflowerlike on top, while stratus clouds spread out in relatively uniform layers. Of course, in a well-developed weather system there may be infinite combinations of these two basic types. (Cirrus clouds, those fine, wispy brush strokes against the blue sky, are of no concern to low-altitude pilots except as precursors of approaching weather. Cirrus clouds are always made up of ice crystals, and are found only at very high altitudes.)

The presence of cumulus clouds always implies vertical currents in the air, very graphically portrayed by the "chimneys" of cumulus clouds in the summer sky; local solar heating creates updrafts that lift warm, moist air to an altitude at which the dew point is attained, and a cloud forms. Every cumulus cloud is the visible cap of a column of rising air; as often as not, the bases of these summertime clouds are very flat and uniform, indicating clearly the altitude at which the dew point and the temperature have come together.

Almost completely opposite in nature, the stratus cloud shares only the principle of formation with its cumulus coun-

Meteorology for the Recreational Pilot

Cloud bases indicate the level at which the air temperature has cooled to its dew point

terpart; when stratus clouds form, you can count on a *lack* of vertical movement in the atmosphere. Stratus clouds are usually formed as a result of cooling from below; the gray, flat skies of winter provide a good example.

Fog is always a stratus cloud. It's called fog only because it forms close to the ground.

As a general observation, the conditions that produce cumulus clouds are more favorable for Recreational Pilots than those situations that result in stratus formations. The vertical currents associated with cumulus clouds usually carry most of the atmospheric debris (haze, smog, smoke) upward, distributing it over a wider and higher area. When there's little vertical movement in the air, the restrictions to visibility are contained near the surface and probably will spread out over a larger area for a longer period of time.

PRECIPITATION

Just because a cloud forms in the sky doesn't mean that there will be precipitation, but there can be no rain or snow or sleet or hail without a cloud to start the process. The cloud requirements—moisture and a small temperature–dew point spread—must precede precipitation of any kind. No matter in what form it finally arrives on the ground, precipitation begins as microscopic water droplets that form a cloud. If the droplets get no bigger, the cloud won't precipitate. But introduce some movement inside the cloud and drops of water collide, each one adding to the weight of the larger droplet, until at last it's too heavy to float and falls out of the cloud as rain. When the air is very cold, water vapor actually condenses in solid form and becomes snow; rain that freezes on the way down arrives as ice pellets; and droplets that bounce around inside a freezing cumulus cloud add layer upon layer of ice until they fall out as hail.

The type of cloud in which condensation occurs usually determines the type of precipitation. Cumulus clouds, in which there is considerable vertical movement and therefore considerable collision of water droplets, can be counted on to produce large-drop showers—the kind that really rattle on the windshield. Conversely, stratus clouds and their characteristic lack of intracloud movement produce misty, drizzly rain with very small drops.

Precipitation in and of itself does not represent a hazard to the Recreational Pilot, but the restriction to visibility that it produces means that when rain or snow is evident in the area you propose to fly in, you must consider the possibility of conditions falling below that 3-mile visibility figure that governs your flight operations. Precipitation forms a sort of veil that prevents you from seeing what might lie on the other side; you cannot take the chance that things will get better before they get worse. As a general operational rule, always fly around precipitation; if that appears impossible, go back, land, and do something other than attempt to penetrate the veil.

WIND

If it were possible to bring the earth's rotation to a halt and shut off the sun for a while, the atmosphere would settle down into an equal-depth layer of air. But with the sun doing its work, the atmosphere would soon develop hot spots, causing the air to expand and rise. Since the atmosphere is fluid, more air would move in to fill the void. That movement is *wind*. A forest fire makes a good, if extreme, example of this process.

The patterns of atmospheric pressure created by uneven solar heating are constantly shifting, and the earth's winds blow in response to these changes; always moving air from high pressure to low, and with greater strength where the pressure differences are large.

Whenever you look at a weather chart that shows highs and lows on the surface, you can count on two things: wind will be blowing clockwise around a high, counterclockwise around a low; and the strength of the wind will be in direct proportion to the strength (measured in atmospheric pressure) of the high or low.

The clockwise-counterclockwise principle is bent just a bit in the nap of the earth, where surface friction tends to slow the wind and change its direction somewhat. This ef-

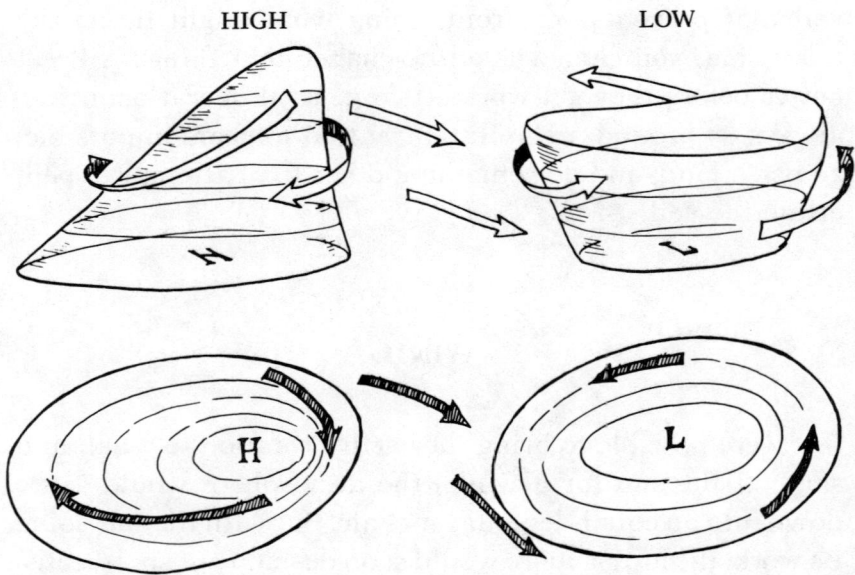

Circulation of air (wind) around and between
high and low pressure systems

fect is generally noticed from the ground up to about 2,000 feet. From there on, relatively free of surface friction, the wind direction increases in bearing (you'd have to turn right to keep the wind in your face) and velocity. As the altitude increases, the general circulation pattern asserts itself even more; it's exceptional to have anything but westerly winds over the United States at 10,000 feet or more above sea level.

Because your flying will be done in a very localized area, it becomes important for you to recognize the conditions that are likely to produce significant winds ("significant" means wind direction as well as velocity, in consideration of crosswind limitations—both yours and your airplane's). Count on strong winds whenever a weather system (a front) is approaching your area, or whenever thunderstorms are rumbling within 10 miles or so. In any event, develop reliable indicators of wind direction and velocity for your aeronautical backyard, and when the speed or direction begin to encroach on your limits, leave the airplane tied to the ground; there will be a better day. Your Flight Instructor will probably make you very aware of limiting wind conditions during training, and will help you work toward safe operations in more demanding wind situations after you receive your certificate.

Just as a good sailor respects and accommodates the force of the wind, a good pilot is able to handle his airplane in conditions up to the level of his skill and experience—and within the limitations of his airplane. Often, the mark of a good sailor or pilot is his recognition of wind conditions that exceed what he or his vehicle can tolerate, and his refusal to sail or fly that day.

There will be precious few days when there's no wind at all. Wind is a fact of life for aviators; it goes with the territory. Whether it's trying to blow your airplane off the runway or creating uncomfortable turbulence in flight, you must learn to counter the wind's effects, to make your airplane do what you want it to in spite of the wind.

FOG

Fog is nothing more or less than a stratus cloud in contact with or very close to the ground, but the word *fog* in a weather report is usually sufficient to send Recreational Pilots home to wait for a better day. Fog is usually completely devoid of turbulence, and seldom extends very far vertically;

the big problem with fog is its restriction of visibility—as often as not, it's a *total* restriction, and always *very* close to the ground.

Fog will not form until and unless the air temperature approaches its dew point, precisely the same conditions required to produce any sort of a cloud. A temperature–dew point spread of 3 to 4 degrees Fahrenheit is considered the brink of fog formation; it's almost assured if there's a light breeze—not more than 10 knots or so—to move the cooling air about. The breeze not only helps create fog, but increases the depth and breadth of its coverage.

You can do a passable job of predicting the formation of fog by being aware of the moisture in the atmosphere (high humidity is the best indicator), and any process that will cool the air. For instance, when the sun sets and the earth begins to cool, the air near the ground is cooled along with it; if the air temperature approaches its dew point—fog. When moist air drifts over cool land areas, such as frequently happens near bodies of water, the temperature drop may be sufficient to generate dense fog.

FRONTS

An "air mass" may be described as a portion of the earth's atmosphere that is homogeneous with regard to moisture, temperature, and pressure. The most obvious quality of air masses is temperature, and so they are usually tagged "cold" or "warm." When the weather people are able to identify a boundary line between two air masses of clearly different qualities, they call the line a "front," and further classify it according to the air mass that is pushing it along.

A cold front can be visualized as the leading edge of a huge bubble of cold air resting on the surface (much like a drop of water on a newly waxed floor), and a warm front is the leading edge of the retreating warmer air.

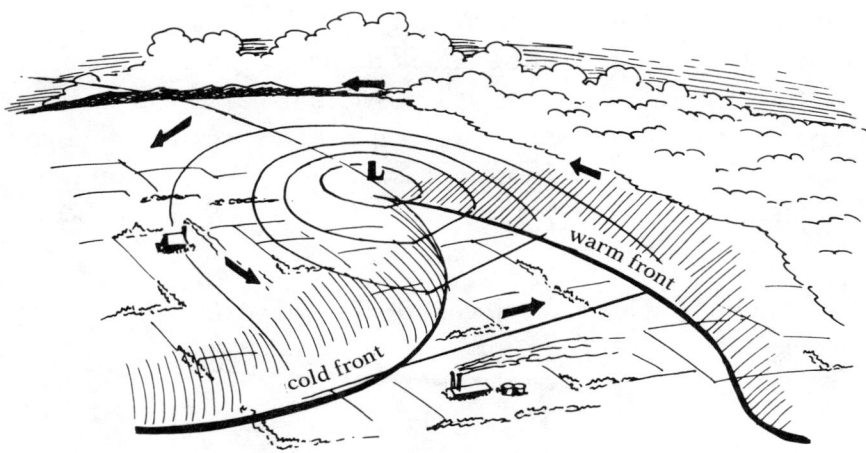

A typical frontal system, rotating around a low-pressure center as the entire system moves generally eastward

As air masses advance and retreat, the weather produced along the fronts can vary from disastrous to nothing at all. Changes in temperature, pressure, and humidity alter and affect the air on both sides of this meteorological war zone (that's why they're called "fronts") and generate clouds, fog, precipitation, high winds—all the forms of weather known to man. In addition, there are some mechanical considerations involved with the movement of a front, because a bubble of high-pressure air muscles other air masses out of the way, with some of the displaced air inevitably being forced upward, expanding, cooling, forming clouds if there's enough moisture present.

The Cold Front and Its Weather

A cold front usually produces a relatively narrow band of weather, often a line of towering cumulus clouds where the unstable warm air is bulldozed upward by the leading edge of the cold air mass. Although cloudiness is part and parcel of most frontal situations, it's possible for a "dry front" to develop, with wind shift and temperature-pressure change the only indicators of frontal passage. The precipitation that almost always accompanies a cold front is likely to be local-

Cross section of typical cold front weather conditions

ized and showery in nature; large drops, heavy rain, or snow that is over within a short while. As the front passes, the drop in temperature is often dramatic, and the pressure may fluctuate rapidly.

Southwest winds precede a cold front, generating considerable turbulence at all levels; the wind speed ahead of the front is often a good indicator of the severity of the weather following not too far behind. Because of the bubble of high-pressure air that generated the cold front, you should expect rapid clearing conditions when the front has passed.

The Warm Front and Its Weather

Warm-front weather is very nearly opposite in character. A wide belt of passive, gentle weather conditions, it's often so wide that the front itself is hard to find. The cooling of moist air as it climbs slowly up the shallow slope of the front forms clouds with little vertical development, and the widespread, small-droplet precipitation that is the hallmark of the warm front results.

Cross section of the weather conditions produced by a typical warm front

Rather than being the abrupt change from hot to cold that is often the case with a cold front, the temperature difference across a warm front is more likely to be characterized as a change from cool to warm, and it doesn't take place rapidly.

Winds ahead of a warm front are much gentler, generally from the south or southeast. Given the wide band of weather common to warm fronts, the clearing of the air after frontal passage is very slow and gradual.

Other Types of Weather Fronts

When a cold front overtakes and overruns a warm front, the meteorologists refer to the result as an occluded front, and it produces a mixture of the weather phenomena that characterize the other two types. There may be locally heavy precipitation inherited from the cumulus development of the cold front, and far-flung stratus and low visibility conditions as a result of the warm front's influence. An occlusion generates its worst weather during the initial stages; as time goes by, energy is dissipated, and the eventual mixing of the two air masses results in a dissolution of the frontal zone.

Both cold and warm fronts are dynamic; they produce their particular brands of weather in part because of the movement across the ground. But when either of them comes to a halt, it is labeled a stationary front, and the weather is much like that of a warm front but less severe. The band of predominantly stratus clouds and drizzle spreads out over a wide area, perhaps moving a bit north during the day in response to the pressure of sun-heated air, then moving southward when the sun sets and cooler air exerts the greater force. A persistent stationary front can vacillate back and forth for days on end, with low ceilings and visibilities over many thousands of square miles. The stationary front is to be expected frequently in the spring and fall, when the battle of cold northern air with warm tropical air is a meteorological standoff. This equalization of atmospheric pressures tends to take place in midcontinent.

AIR-MASS WEATHER

After a front passes, weather conditions depend mostly on the moisture content of the air mass and the heating or cooling that may occur. A moist air mass over land of a lower temperature will be cooled from below, and you should expect low ceilings and visibilities, the degree of which will be a function of the dew point–temperature relationship. On the other hand, a moist but clear postfrontal air mass that is heated by the surface will likely produce localized cumulus clouds—"fair-weather cumulus"—and scattered thunderstorms on occasion. When an air mass is very dry, look forward to fine, bright weather on the west side of the front.

When an air mass comes to a standstill, it begins to take on the characteristics of temperature and humidity that exist on the surface. Significant modification can occur when an air mass remains over a particular land or water area for an extended period of time; moisture content may increase, temperature may rise or fall and provide a completely different set of results when the air mass is finally pushed out of the way by the next weather system.

THUNDERSTORMS

Aviation educators have for so long taught and warned of the horrible things that can happen to pilots in and near thunderstorms that the question must be asked: "Are thunderstorms really as bad as they say?" And the answer is an unequivocal YES—and then some. Your 3-mile visibility limitation will certainly keep you from flying into one, but there are hazards that exist outside the storm, and you have no choice but to give *any* thunderstorm a very wide berth. They are awesome, unpredictable, intimidating, and *dangerous*.

Thunderstorms can be expected to occur whenever the air is sufficiently unstable and some sort of lifting action takes place. The steep slope of a rapidly moving cold front is a prime producer of the lifting action that triggers heavy thunderstorms in advance of the front; rising terrain is another culprit. Nearly all year long in the south, and during the summer months in most of the rest of the United States, uneven heating of the earth by the sun creates convective currents, which lift unstable air thermally and set off thunderstorms.

The growth process of a thunderstorm is predictable and usually visible, and occurs in three stages: cumulus (building), mature, and dissipating.

In the cumulus stage, unstable air has been lifted and a cloud with vertical development forms; all air movement is upward, and precipitation has not yet started. Sooner or later, the droplets of water inside the cloud become so large and heavy that they can no longer be borne upward by the vertical currents of air, and rain begins, dragging air along with it.

The line between the building stage and full maturity is a very thin one, but when the storm is fully developed there are airplane-bending combinations of adjacent updrafts and downdrafts inside the cloud, heavy precipitation under the storm, and great heavings of the atmosphere that may dis-

Meteorology for the Recreational Pilot

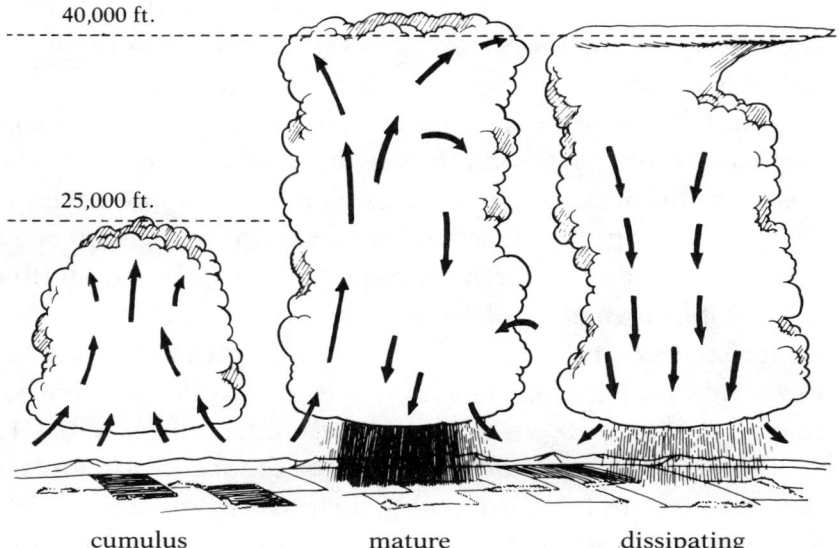

Three stages in the life cycle of a typical thunderstorm

turb the air for miles around. The mature stage is universally recognized as the most dangerous; all the hazards will be at their worst when this condition is attained.

The dissipating stage of a thunderstorm's life cycle begins when the energy inputs slow down. Perhaps the cold front slows, until the lifting action is insufficient to keep things going, or the sun sets and removes the heat source that started the whole process, or the storm itself moves with prevailing winds and drifts away from the hot spot.

Hazards outside the storm—those of greatest concern to the Recreational Pilot—include high speed, gusty surface winds that can change direction rapidly; heavy turbulence in the clear air for several miles around the storm; and heavy rain, which lowers visibility and frequently masks other storm cells. Hail is produced by most thunderstorms and, if thrown out of the top of the cloud, can fall in the clear air downwind of the storm proper. Flying into a flock of ice cubes is a surprise at best, a disaster at worst.

A tall, broad thunderstorm (height is the best indicator of intensity) shuts out some of the light of day, hastens the onset of darkness, and causes problems for the pilot who is determined to get home before darkness arrives. On top of all these unpleasant features of thunderstorms, the possibility of invisible tornadic tubes (now known to circulate inside major storms, and suspected of existing in the clear air *between* adjacent thunderheads) makes it easy to advise Recreational Pilots about flying near thunderstorms: DON'T!

Because the conditions required for thunderstorm formation are so simplistic—unstable air plus lifting action—it's not difficult to predict when they are likely to occur. Any time the local atmosphere is warm and moist, the potential has been halfway satisfied; add some kind of lifting, and cumulus activity will result. *Every* approaching cold front should be immediately suspect, with the fast-moving ones deserving even more attention; they often exert their influence hundreds of miles ahead, in the form of an instability line (much like the bow wave of a boat), which generates *squall line thunderstorms*, the most violent of them all.

HIGH WINDS

The lighter the aircraft, the harder it is to overcome the effects of the wind, and the greater the hazard of high winds—at a sustained velocity or in the form of gusts. In addition to the obvious problems of aircraft control when

Meteorology for the Recreational Pilot

the wind is blowing faster than the stall speed, the vertical differential in wind speeds creates its own special characteristics. If air moved rapidly without stumbling there would be less of a problem, but whenever the wind blows faster than 15 knots or so, it is slowed significantly by contact with the ground; expect a sudden increase in wind velocity right after takeoff, and an equally sudden decrease as you approach the runway for landing. Sailplane pilots compensate by approaching at a higher airspeed; power pilots have additional thrust available to make up for the sudden loss of performance.

Gusty winds are uncomfortable and distracting, but the big problem is the rapid change in the angle of attack when a vertical gust in encountered. Given the normal relative wind, parallel and opposite to the flight path of the airplane, a gust of wind that is directed even slightly upward will subject the wings to an immediate, though short-lived, high angle of attack. Carried to extreme, the momentary high angle could cause the wing to stall; if this should happen close to the ground, a very hard landing (or worse!) might result. A prudent pilot flies at a slightly higher airspeed when the final approach path is populated by gusty winds.

WEATHER REPORTS

Flight Service Stations (FSS) are operated by the Federal Aviation Administration just for pilots, and are accessible from any telephone (you'll find toll-free 800 numbers listed for this purpose). The FSS is the fountain of knowledge for Recreational Pilots in the matter of weather information; a

Recreational Flying

personal visit or a telephone call provides immediate access to the world's most extensive weather-information system, and to a specialist trained to help you interpret whatever reports and forecasts might apply to the flight operation you propose. In those areas where the FSS gets far more telephone requests for weather information than it can possibly handle, special numbers are listed that provide a continuous, updated recording of the briefings for popular flight routes in that area.

All of the weather data in the national system is yours for the asking, and although a Recreational Pilot could be overwhelmed in a hurry by a full preflight briefing, there are certain fundamental items of information that can be most helpful.

When the FSS specialist answers, identify yourself as a pilot and provide your airplane's "N" number. Of primary interest should be the general forecast for the local area, specifically the presence of fronts or other conditions that may produce low visibilities, low ceilings, high winds, and the like. Keep in mind that the weather observations taken each hour at major airports contain information about ceiling height, visibility, temperature, dew point, and wind conditions.

The FSS can provide reasonably accurate forecasts for a twenty-four-hour period, so that you can plan tomorrow's flight activities today. Forecasts are available for specific airports (if there's one of these close by your home base, the weather will probably not be much different where you intend to fly), and for much larger areas, usually encompassing several states. The standard contents of this forecast deal with the position and expected movement of weather systems, cloud bases and tops, visibility over the area, and turbulence, among other things.

You will need to develop a set of reliable visibility checkpoints around your home airport, so that you can tell at a glance when your 3-mile visibility limit has been reached. Prominent buildings, towers, hills, and so forth provide good references; pick a series of checkpoints around a circle, because it's not all that unusual for the visibility to be very good in one direction and very bad in another.

The bases of the clouds (ceiling) must be of concern, although the Recreational Pilot rules contain no reference to altitude limitations. That's because you must stay at least 500 feet above the ground except during takeoff and landing, and since it doesn't make much safety sense to fly around at 500 feet, any solid cloud cover lower than 1,000 feet should get your attention. A good way to develop the ability to estimate the height of cloud bases is to compare your observations with the official report from a nearby airport; before long, you'll be able to tell whether you should attempt a flight or not.

In addition to the official government sources, the daily newspaper and the television set offer weather information that is very general in nature and coverage, but can at least give you an insight into the kind of weather you should expect today or tomorrow or the rest of the week. Newspapers publish a facsimile of the national forecast that shows fronts, precipitation, the positions of highs and lows—enough to give a knowledgeable pilot a good idea of the conditions coming up.

TV weather shows at the local level vary from excellent to hardly worth watching, but there is a pilot-oriented weather show on the Public Broadcasting System each weekday morning—"AM Weather"—which is definitely worthwhile. The Weather Channel, a cable TV presentation, is a twenty-four-hour-a-day program with outstanding information for pilots.

LOCAL WEATHER PHENOMENA

All of these weather situations are the classics, the expectations of the weather prophets when things happen according to the book. But just as likely as not, the modification of an air mass or the route of a major system will be altered remarkably by the terrain, a large body of water, a heavy concentration of industrial activity, or any one of a countless number of combinations.

Because you'll be flying locally most of the time, it's a good idea to "keep book" on the weather as it occurs, noting the conditions that tend to set up the circumstances that prohibit or limit your flying. The next time the same set of precursors shows up, you'll be one step ahead of the game.

RECOGNIZING CRITICAL WEATHER CONDITIONS

By the time you complete your training for the Recreational Pilot certificate, including studying for the written examination, you will have begun to develop a storehouse of personal weather observations related to the feasibility and safety of flying. You will probably experience—in the company of your Flight Instructor—some wind conditions that push the limits of your ability and your airplane's limitations; you will have flown on days when the visibility is approaching three miles, and will be able to know just from looking that this is a weather condition in which you shouldn't be aloft.

On a day when the cloud cover obviously presents an upper limit to your flying, keep track of your altitude above the ground, and when the clouds force you to fly uncomfortably close to the ground—say 1,500 feet or so—give strong consideration to calling it off for the day. The same thinking should apply to your visibility observations from the cockpit; when haze or rain or fog begin to limit how far you can see, and you know that flight visibility is down to 5 miles or so, head for the barn. A conservative approach to the weather has saved a lot of pilots a lot of embarrassment, or worse.

Scattered clouds becoming solid layers above or below you, dark areas showing up in the clouds, precipitation, and evidence of increasing wind speed (flags, smoke, waves), are all signs of impending weather conditions that may be beyond your limits. When in doubt, get on the ground, and come back to fly happily and safely another day.

Keep in mind that the Recreational Pilot certificate was created to allow pilots to fly for *fun*; high winds, turbulence, fog, and rain present weather conditions that run counter to that philosophy, and you'll do better to stay on the ground. The rewards of flying on beautiful sunny days far overshadow the occasional inconvenience of having to cancel a flight when the weather is less than ideal.

17

Aircraft Loading

ALL AIRCRAFT—airplanes, helicopters, blimps, gliders, hot-air balloons, kites—are sensitive to the loads they carry. In some cases, it's a matter of simply being too heavy to get off the ground; in others, sheer weight coupled with the distribution of that weight can have a significant effect on handling qualities and flight characteristics.

Fortunately, the airplanes flown by Recreational Pilots are not likely to suffer from serious loading problems. These airplanes have limited seating capacities (to say nothing of the one-passenger limitation), and for the most part the fuel tanks and baggage compartments are located very close to the center of gravity. But you must still consider proper loading procedures, which means that you must understand the principles, legal limitations, and be able to perform the calculations that prove everything's okay. Federal Aviation Regulation 91.3 puts the overall responsibility for the safety of the flight on your shoulders, and FAR 91.31 requires that you do whatever is necessary to stay within your airplane's operating limitations, of which the load limitations are a significant part.

Aircraft Loading

WEIGHT CONSIDERATIONS

The legal limit on aircraft weight is directed more toward performance and maintenance of structural integrity; in other words, keeping all the pieces of the airplane together in flight. An overloaded automobile may groan across the ground or collapse in a heap of broken springs, but an overloaded airplane can be put into a compromising situation when, for example, it's too heavy to fly and is moving too fast to be stopped on the runway. Occasionally there's enough lift to get the overloaded airplane off the ground but it refuses to climb. If the engine is putting out 100 percent power and the wings are generating all the lift they can and it's still not enough to clear whatever's in the way, the result is at best embarrassing, at worst tragic; and it can be traced right back to an easily committed sin—the pilot tried to fly an overloaded airplane.

G force, or load factor: the effect of centrifugal force in a turn or a pullout from a dive

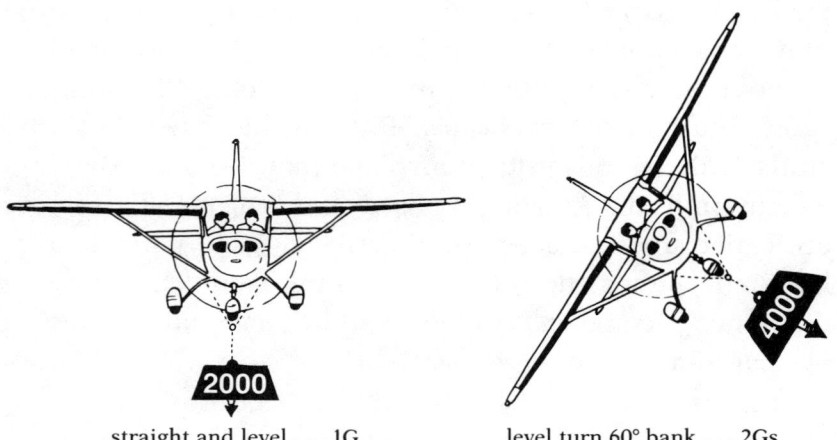

straight and level . . . 1G   level turn 60° bank . . . 2Gs

Recreational Flying

Beyond performance considerations, airplanes suffer from yet another limitation: *load factor*. An airplane is frequently subjected to centrifugal force, which effectively increases its weight. For example, if you should roll your airplane into a 60-degree bank and apply enough back pressure on the controls to maintain altitude, the centrifugal force exerted will be precisely the same as the actual weight of the airplane, and will add to the load the wings must carry. Your 2,000-pound airplane suddenly weighs *4,000* pounds as far as the wings are concerned. The *effective weight* in such a turn (or any other maneuver that imposes a similar load) is twice the actual weight, or a load factor of 2. You'll also see this force expressed in terms of units of gravity—G—because any load at rest weighs whatever it weighs, with only the pull of gravity (l G) acting on it. Whenever the load factor increases, the G-loading will increase with it, so a load factor of 2 is the same as 2 Gs on the airplane and everything in it.

Except for those purposely designed for other than normal passenger-carrying chores, airplanes are built to withstand a load factor of 3.8, so you can wrestle that 2,000-pounder around the sky, turn as sharply as you like, even play dive-bomber—the wings will stay with you up to an effective weight of 7,600 pounds. Beyond that, the manufacturer's guarantee evaporates. But consider the pilot who stuffs 2,200 pounds into an airplane that has a 2,000-pound weight limit. If 3.8 Gs are imposed on this airplane, its wings are called upon to support, effectively, 8,360 pounds. Two hundred pounds the pilot didn't think would make much difference become 760 pounds of additional load. The lesson is clear and simple: *don't overload!*

Aircraft Loading

IMPORTANT LOADING TERMS DEFINED

If an airplane were put on the scales just prior to takeoff, the dial would show the weight of the airplane itself plus the passengers, baggage, and fuel; the pilot is responsible to see that this weight does not exceed the *maximum allowable gross weight*. Each airplane within a model designation (such as a Cherokee 140 or a Cessna 152) has the same maximum weight limit, because the structural considerations are identical, but the *empty weight* of the airplane (just the airplane itself, no people, baggage, or fuel on board) varies with the owner's preference for optional equipment and subtle differences in the manufacturing process. The empty weight is the starting point for your loading calculations, the number to which you add pounds of people and baggage and fuel. Empty weight is so important that it's specified in each individual airplane's papers, and must be officially recomputed whenever it changes.

Between the empty weight and the maximum gross weight is the number of pounds you have to work with: *useful load*. You can put together any combination of occupants, baggage, and fuel, but the useful load may never push the total weight over the top; you will always and ever be limited to the maximum allowable gross weight at takeoff.

In practice, you need be concerned about only one of those three numbers. The empty weight is fixed, the maximum allowable gross weight can't be exceeded, so if you're careful never to put on board more than the useful load, you can't get in trouble weightwise. When you have doubts—even little ones—about the legality of a particular load, total up the weights and make sure that what you propose putting into the airplane will fit between the empty weight and maximum gross weight.

Specific restrictions on loading will show up in many airplanes; such limits are usually indicated by placards, and are always included in the operating limitations. Typical are those on baggage compartments and hat shelves, where the structure can support only a certain number of pounds when the 3.8 load factor is applied. A shelf limited to 20 pounds, for example, is actually 3.8 times stronger than that, but more than 20 pounds of hats could cause structural failure if more than 3.8 Gs were imposed on the airplane.

PLACING THE POUNDS PROPERLY

Three items are involved in understanding, preventing, and correcting the problems of in-flight balance: center of lift, center of gravity, and the aerodynamic power of the elevator to control nose-up and nose-down movements.

The elevator exerts up or down force depending on its position (streamlined or deflected into the airstream) and the speed of the air flowing across it, with a near-infinite range of elevator control force available, from zero when there's no movement of air, to a great deal when the airplane is flying at a high speed. The *center of gravity* (CG) is the point at which all the weight of the airplane appears to be concentrated. The *center of lift* (CL) is the point at which all the lift appears to be pushing upward; it moves back and forth somewhat in different conditions of flight, but for simplicity's sake, consider it as acting upward right in the middle of the wing.

Aircraft Loading

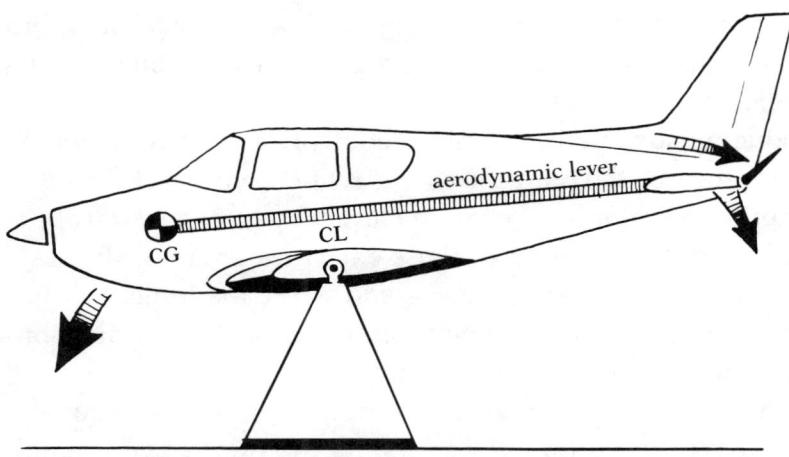

forward center of gravity = nose heavy

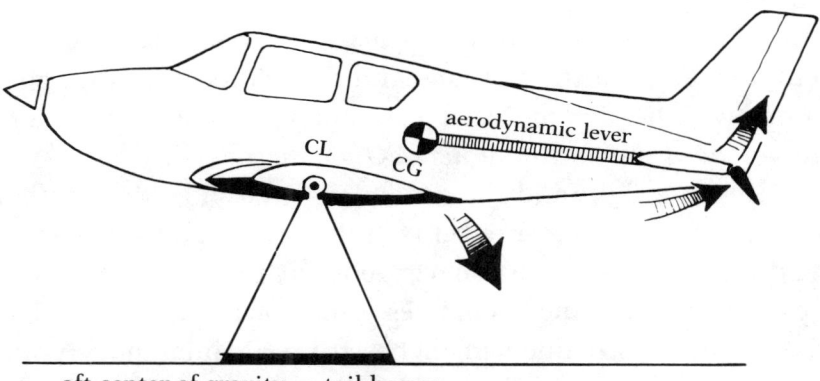

aft center of gravity = tail heavy

Relationship of the center of lift, the center of gravity, and the aerodynamic power of the elevator

Recreational Flying

The CG positions shown (page 195) are exaggerated, but you can see that a forward CG provides a very long aerodynamic lever or arm through which the elevator force can act, producing supersensitive pitch control. If the CG is located too far toward the rear, the arm is reduced considerably, and the pilot may not have enough elevator power to control the airplane properly. Both of these extremes are undesirable—even dangerous in some cases—and so the manufacturer has arrived at a compromise between reasonable load distribution and handling characteristics.

Every airplane therefore has an acceptable *range* for the CG, with forward and aft limits. As the pilot in command, you are responsible to make certain that the airplane is loaded so that the maximum weight is not exceeded, and that the center of gravity is located within the acceptable range. You'll need to go through a simple arithmetic exercise to accomplish this.

All loading computations start from the same base—the empty weight of the airplane and the CG location that goes along with it—and arrive at a common point of information: *total loaded weight* and a *new CG location*.

Empty weight and the center of gravity location in that configuration are required entries in the airplane documents, and must be revised whenever modifications or addition or removal of equipment changes either of the numbers. The CG location is defined in inches from an imaginary point close to the nose of the airplane; this point is called the *datum*.

Suppose a certain airplane weighs 1,000 pounds in the empty weight configuration and its center of gravity is located 100 inches from the datum. If all 1,000 pounds could be bundled together and placed at that point, the downward force would be defined as a *moment* of 100,000 pound-inches,

Aircraft Loading

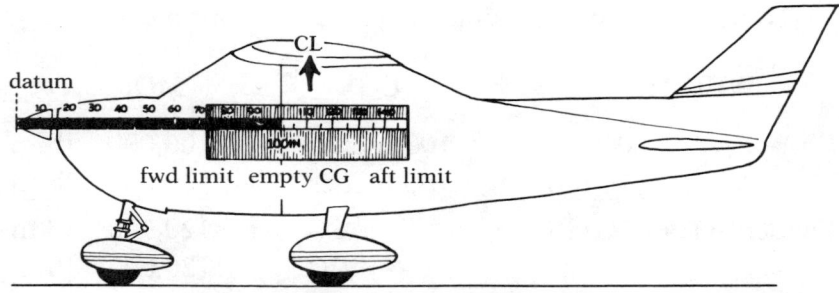

EMPTY AIRCRAFT WEIGHT 1,000 LBS.

Empty weight and the corresponding CG location are the foundations on which every airplane loading problem begins

or the force of 1,000 pounds at the end of a lever 100 inches long. It's an imaginary lever and an imaginary concentration of weight, but by considering airplane loads in these terms, the effect of any given weight can be determined.

From the base numbers of empty weight and CG location, it's a very simple exercise to add up the weights of people and baggage and fuel, *and* the effects of each one of those items in terms of moments. You need to know how much each load item weighs, and where it will be placed in the airplane. For example, a 200-pound pilot would create a moment of 20,000 pound-inches if the front seat were right over the CG (200 pounds acting through an imaginary lever 100 inches long).

When the weights and the moments are totaled, the new center of gravity is determined by solving this relationship:

WEIGHT times CENTER OF GRAVITY equals MOMENT

therefore

$$\text{CENTER OF GRAVITY} = \frac{\text{MOMENT}}{\text{WEIGHT}}$$

Recreational Flying

To prove the point using the previous example,

WEIGHT	×	CG	=	MOMENT
Empty wt 1,000 lb	×	100 inches	=	100,000 lb-in
Pilot 200 lb	×	100 inches	=	20,000 lb-in
Loaded wt 1,200 lb	×	CG	=	120,000 lb-in
		CG	=	$\dfrac{120,000 \text{ lb-in}}{1,200}$
		CG	=	100 inches

With the pilot's seat at the CG, nothing changes but the weight. A passenger in the right front seat will add only weight—no shift in the center of gravity. Anything loaded in the rear seats or the baggage compartment will move the CG to the gear.

There are several methods of calculating load distribution; the first is a continuation of the problem above. Multiply the weight of each item you put on the airplane by its arm (the number of inches from the datum, which information is always supplied by the manufacturer) to get the total weight and total moment. Make sure the gross weight is no more than is allowed, then divide the moment (always the larger number) by the weight to arrive at the new center of gravity location in inches from the datum.

Certain manufacturers provide a loading graph, which does the multiplication for you, and some airplanes will have a table that simply provides a permissible range of moments for a number of airplane weights; when the loaded CG (expressed in pound-inches) falls between the minimum and maximum moments on the table, the load is within limits.

Whether you get there by means of simple arithmetic or a plastic plotter, there's some kind of document for every

Aircraft Loading

airplane that provides center of gravity limits (fore and aft) at various weights. Your task is merely to check the numbers to be sure they're within those limits. If they are, go flying, secure in the knowledge that the airplane will perform safely and predictably.

18

Aircraft Peformance

A PILOT who flies the same airplane time after time will begin to recognize its limitations; he'll know when a runway is too short or a mountain too high. But the time-after-time process that leads to that knowledge can be peppered with some pretty harsh lessons. Learning how much runway an airplane requires by plunging off the end of one that's too short is not a good deal in anyone's book, nor is the uncomfortable realization that the hill ahead can't be flown over. Experience is the best teacher, but it's also the most expensive.

The length of the runway required for takeoff and landing, the rate at which the airplane climbs, and the speed at which it cruises are predictable, reliable numbers which represent performance. In this chapter, we'll discuss *normal* performance, the sort of behavior you should count on in everyday flight operations; in chapter 19, the emphasis shifts to *maximum* performance maneuvers, for those times when the situation requires every bit of hustle your airplane can produce.

Aircraft Performance

DENSITY ALTITUDE

The performance of all flying machines that depend on air for thrust and lift is very significantly affected by the density of that air. No matter how good its pilot or how impressive the sales-brochure statistics, an airplane's performance improves in thick air and gets worse as the air gets thinner. The range of air density at any specific location (such as the airport from which you operate most of the time) is not usually a very wide one, and since one extreme—a very dense-air situation—can only improve performance and sweeten your flying experience, your concern should be to recognize and take into account the effects of low-density air.

Certain immutable laws of physics dictate the responses of a gas (such as the atmosphere) to changes in temperature and pressure. If the gas is uncontained—that is, free to expand and contract—it will increase its volume when the temperature is increased or the pressure is decreased. In very general terms, you should expect lousy airplane performance on a day when the temperature is high and the pressure is low.

Temperature variations are easy to understand and recognize, but changes in pressure are not so easy to detect. Very seldom does the barometer venture more than a half inch above or below the normal 29.92 inches of mercury, and we have gotten so accustomed to these slight, slow changes in pressure that they aren't usually noticed. On top of that, the air pressure changes that change airplane performance come from two sources: the weight of the atmosphere pressing against the surface of the earth and changing constantly as weather systems move along, and the altitude of the airplane. After all, we live on the floor of an ocean of air, which means that each foot the airplane climbs decreases the pressure a little bit.

Rather than calculate the effects of pressure and temperature on each contributor—engine power, thrust, and lift—the aviation community puts them all together and furnishes performance information for a standard situation, with corrections for abnormal atmospheric conditions. The *standard* situation is the pressure-temperature combination at sea level when the barometer reads 29.92 inches of mercury and the temperature stands at 15 degrees Celsius. Under those conditions, air density is constant and predictable; if standard-day conditions prevail throughout the atmosphere, each 1,000 feet of altitude increase will result in a decrease of 1 inch of pressure, and a drop of 2 degrees in temperature.

Given those standards, air density can be calculated for any altitude. For example: at a sea-level airport on a standard day, the pressure is 29.92 inches, the temperature is 15 degrees, and the altitude is 0 feet; at 5,000 feet above that airport, the pressure should be 24.92 inches (five times the lapse rate of 1 inch per 1,000 feet) and the temperature should be 5 degrees (five times the lapse rate of 2 degrees per 1,000 feet). The air would be as dense as 5,000-foot air *should* be on a standard day, and your airplane would perform accordingly.

But suppose that because of a particular weather system the air pressure drops more rapidly as you climb, and the temperature doesn't drop as fast, and when you reach 5,000 feet the air is much less dense than it should be. If, for example, the air density is the same as it would be at 6,500 feet on a standard day, your airplane will perform *as if it were flying at 6,500 feet*. Climb rates and cruise speed will be lower than normal, and if you should land or take off at an airport located at that altitude, you'd find that considerably more

Aircraft Performance

runway would be required. In this example, "actual" altitude is 5,000 feet, but the *density altitude* is 6,500 feet—and that's the altitude your airplane senses.

Airplane manufacturers have long been required to supply a complete set of performance charts with each airplane, but they have not always paid attention to standardization of these charts, for each has his own method of presenting the information. At any rate, there will be a series of charts or tables which will allow you to figure out just how much runway you'll need on a given day, how fast the airplane will climb, how fast it will go in level flight, and how much fuel it will burn to make all that happen.

WEIGHT AND WIND

No matter which type of chart or table you use, there are things other than density altitude to be considered, such as airplane weight, which will obviously cut down overall performance as the number of pounds on board increases. On takeoff, for example, the wings require a certain airspeed to generate enough lift, and since thrust will be the same regardless of weight, a heavier airplane will accelerate more slowly and use up more runway in the process of attaining takeoff speed. In cruising flight at a given power setting, a heavier airplane will inevitably fly slower than a lighter one.

Airplane weight really comes into the picture during landing. Suppose that you are on final approach to a short field with a heavy airplane. Even at the slowest possible approach speed, every extra pound on board adds to the mass moving down the runway, a mass that comes to a stop only when all of the energy is somehow dissipated . . . through heat in the brakes, rolling friction, running the airplane through a hangar, or any one of several other effective but unpleasant alternatives.

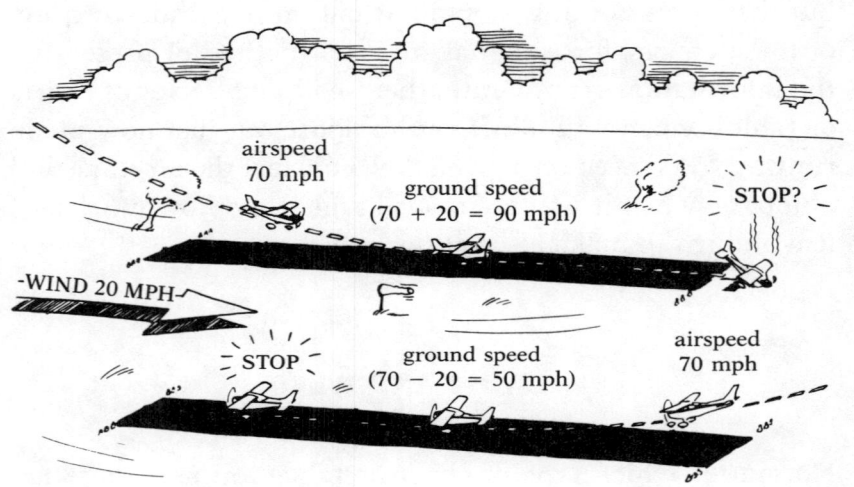

Runway required to bring the airplane to a stop
is always increased by a tailwind

The distance required to become airborne and the distance to touch down and stop are both very dependent on the wind. With all other factors constant, an airplane will get off the ground sooner and consume fewer feet of runway on landing when those two operations are conducted into the wind. On takeoff, it's a matter of the wings' sensing airflow even while they're standing still; airflow means lift, even if only a small amount, and whatever lift is generated by virtue of the wind is that much less to be developed from forward movement of the entire airplane. On the other end of a flight, an into-the-wind landing is a higher-performance landing

Aircraft Performance

because the groundspeed and therefore the kinetic energy are less.

Any air movement from behind (a downwind takeoff) means that your airplane must accelerate to normal takeoff speed plus the speed of the wind before the wings can provide sufficient lift to get everything off the ground; and more acceleration means more distance from brake release to liftoff. Kinetic energy on a downwind landing is increased considerably, because the wind from behind pushes the airplane across the ground, and the stopping distance grows accordingly.

Takeoff and landing performance is always improved by operating into the wind, hence the general rule that pilots leave the earth and arrive thereupon using the runway most nearly aligned with the wind. The airplane's performance charts will correct (improve) the takeoff and landing numbers accordingly, but very seldom will you find a performance chart that takes a tailwind into account. The manufacturers are trying to tell you something; if you haven't enough experience to quantify the effect of a tailwind, consider all tailwinds bad news and make your takeoffs and landings with the wind in your face.

RUNWAY CONSIDERATIONS

The takeoff and landing performance charts might be considered a contract between you and the folks who built the airplane; you agree to supply normal pilot technique, and the manufacturer supplies airplane performance in consideration of all the factors of wind, weather, and weight. But the fine print in that contract usually specifies a level, hard-surface runway. What about takeoff or landing on a sloping airport? Or one of those roller-coaster runways where the middle of the strip is considerably lower (or higher) than either end?

Faced with an even slope and no wind, you can put gravity to work for additional acceleration, so a downhill takeoff is definitely advised. When the wind is blowing down the slope, you'll have to make a judgment on the merits of downhill acceleration versus a tailwind takeoff. In the case of a sunken-center runway, you'll have help to the halfway point, but it's all uphill after that; a humpbacked airport will be just the opposite. You'll find no charts for runway slope in the light-plane performance books, so experience (look to your Flight Instructor for help here) becomes the deciding factor.

Numbers from the performance charts are also invalid when the runway surface is anything except pavement. The wheel drag of even a dry, just-mowed grass runway will increase the takeoff distance somewhat, but let the grass grow a couple of inches, let the surface get soggy from rain, and it is quite possible that the power *available* will fall short of the power *required* for acceleration to takeoff speed. As in the case of runway slope, a pilot judgment will have to be made, and once again it's experience that makes the difference. Long grass, snow, mud, and the like will probably increase the takeoff distance by 100 percent or more. When in doubt, try it with a very light load, or mow the grass, or both!

VERTICAL PERFORMANCE

For any given airplane weight, there is one, and only one, airspeed that will produce the maximum rate of climb. It's a

Aircraft Performance

compromise of vertical and horizontal movement. It's known officially as the *best rate of climb* speed, and it provides the most gain in altitude per unit of time.

**MOST GAIN IN ALTITUDE PER UNIT OF TIME
(FEET PER MINUTE)**

Climbing at the best *rate* of climb airspeed

On the other hand, absolute maximum climb performance over a short distance can be achieved by increasing the angle of attack until the wing is operating just short of a stall, producing all the lift of which it's capable. The increased rate of climb that results is at the expense of airspeed, but when you need to climb over a tree *right now*, the angle of climb is a heck of a lot more important than your progress across the ground.

The *best angle of climb* speed (always a lower number than best rate) will provide the most gain in altitude per unit of distance. *Climb performance—best angle and best rate—presupposes the use of full power in all cases.*

The airspeed for best angle of climb requires a pitch at-

Recreational Flying

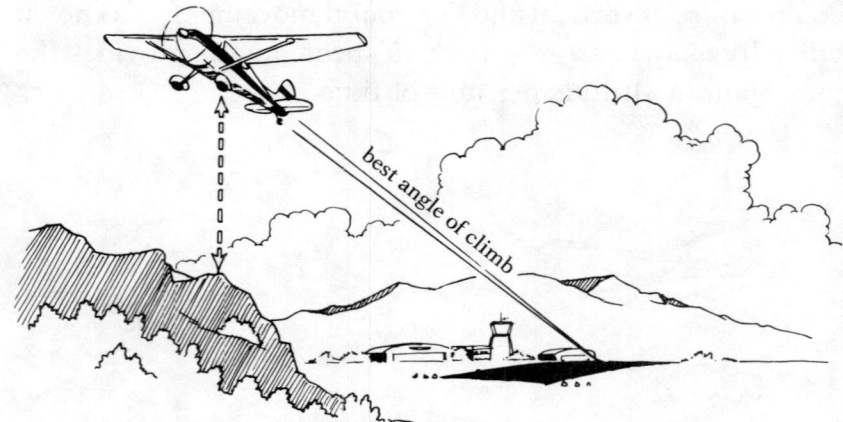

MOST GAIN IN ALTITUDE PER UNIT OF DISTANCE
(FEET PER MILE)

Climbing at the best *angle* of climb airspeed

titude that will fill your forward vision with a great view of the airplane's engine, frustrating your attempts to make sure there's no one else out there ahead of you—and even at the best rate airspeed, you may feel uncomfortable about what you can't see out front. The solution is to either climb at a slightly higher airspeed, lower the nose, or turn frequently during a climb, or all three; do whatever you feel is necessary to clear the area ahead.

While best angle of climb airspeed is most useful for clearing obstacles in the immediate takeoff path, it's a good idea to maintain this speed during the first 500 feet or so during every takeoff. Keep in mind that in the event of an engine failure in a single-engine airplane, the most valuable commodity is *altitude;* it provides more time for maneuvering to a suitable landing site. It makes a lot of sense, therefore, to gain just as much altitude as is possible in the early stages of a flight. As soon as you've got a comfortable 500 feet between you and the ground, it's good practice to lower the nose a bit and continue the climb at the best rate airspeed.

Because it's based on maximum efficiency of the wing, the best rate airspeed finds application at the other end of the airplane performance spectrum as well. Best rate speed turns out to be very close to the *least rate of descent* speed; there will be some slight changes that result from the lack of thrust, but when the engine quits, you can glide farther at that airspeed.

THE MOST IMPORTANT FACTOR: PILOT TECHNIQUE

Density altitude is undeniable, weight is controllable, the runway slope is unchangeable, and wind effects can be determined right up to the moment the brakes are released for takeoff, but the single element that probably has more to do with what actually happens in regard to airplane performance is the one that is least predictable and varies more than all the others combined: pilot technique.

Most performance charts leave a little room around the edges of the numbers to allow for flying skills that may be a bit rusty, or were never really fully developed. But remember that those numbers were obtained with expert pilots at the controls of new airplanes, with new engines, and it's doubtful that the average pilot can achieve all the performance of which his airplane is capable, if for no other reason than that a reciprocating engine starts grinding on its insides and losing horsepower the very first time it's run, and continues downhill from there—a small loss, but a performance thief nonetheless.

Recreational Flying

Given the overwhelming availability of comfortable-length paved runways, and the infrequency of occasions when a Recreational Pilot really needs to fly into or out of a short strip, there are two ways you can keep from painting yourself into a performance corner: if you *must* operate from a runway with little margin for error, get good at flying your airplane under those conditions (that requires training and experience); and when the amount of ground or sky that you'll need comes close to the amount available, remember that it's you versus the test pilot—add something for lack of experience and skill (that requires personal evaluation). And should the situation smack of impossibility, don't be there at all—*that* requires common sense.

19

Maximum Performance Operations

CHANCES are pretty good that most Recreational Pilots will never have the occasion, much less the need, to operate from a short field, one that severely taxes his or her airplane's abilities. Even with this low probability, it's good to know what maximum performance is all about, especially with regard to landings; you never know when you might have to put down in a farm field or a clearing in a forest that isn't long enough to fit into the landing distance you've been using back home on a long, paved runway. In general, the pilot who has some experience in maximum performance operations in a training situation (with no obstacles other than imaginary ones, and plenty of runway ahead to accommodate the inevitable mistakes of a learner) will be able to do a much better job of flying into or out of a short field even if it requires only half the ultimate performance of the airplane.

Recreational Flying

"Maximum performance," just like "normal," needs to be defined: on takeoff, you make the airplane lift off the ground in the shortest possible distance, then climb at the steepest possible angle until all obstacles are cleared; on landing, you clear whatever obstacles lie in the approach path, then bring the airplane to a full stop in the shortest possible distance on the runway. For our purposes, maximum performance also refers to the specialized techniques and procedures required to operate safely when the airfield is soft or rough.

Maximum performance is essentially energy management: making the airplane's capabilities fit the situation at hand. True maximum performance puts airplane and pilot at the very edge of their capabilities; it's advanced stuff, and shouldn't be tried without the guidance of your Flight Instructor.

THE SHORT-FIELD TAKEOFF

The objective is to get the airplane from a standing start to a height of 50 feet as soon as possible. Why 50 feet? Because that's the way it's always been; somebody decided many years ago that this height was right for determining airplane performance—possibly because many early airports were surrounded by 50-foot trees—and the number stuck.

First, check the Pilot's Operating Handbook for the proper flap setting if your airplane is so equipped. Almost all of the airplanes flown by Recreational Pilots use zero flaps for a short-field takeoff, but there are some machines out there in RP land which can generate maximum performance only with partial flap extension; double-check to be sure.

Maximum Performance Operations

Once in position on the runway (it's important to line up at the very beginning of the takeoff surface, to make every foot of the runway available), hold the brakes and run the engine to full power before starting the takeoff run. This is your last chance to check power output, and with the throttle wide open, there's no delay due to power application.

All set? Release the brakes and maintain directional control with the rudder, allowing the elevator to "float" in the airstream. There will be less drag this way, and you need to let the airplane accelerate to takeoff speed just as quickly as possible.

As the airspeed indicator needle moves toward the lift-off speed published in the handbook, begin to rotate the nose upward to an attitude that will stop the indicator in its tracks, and maintain that speed throughout the climb to 50 feet. Don't let the airplane fly off the ground as in a normal takeoff; this is a much more aggressive maneuver, and you'll need to *pull* it off the runway. Practice will help you develop not only the proper rate of rotation, but easily recognized visual clues (rivets, screwheads, and the like) that let you know the nose is where it belongs on the horizon.

After the airplane climbs through 50 feet or the actual height of the obstacle, lower the nose a bit to the attitude for a normal climb (at the best *rate* airspeed), and go about your business.

Recreational Pilots flying tailwheel airplanes should alter this technique slightly. Everything is the same until the airplane starts to move; then, as soon as you feel the elevator come alive, get the tail up. This streamlines the airplane as much as possible, and permits rapid acceleration to lift-off speed. From there on, the procedure is the same.

In either case, use the same technique every time. Even though atmospheric conditions, winds, and airplane weight will change the number of feet required to reach 50 feet, you will be doing everything possible to maximize the airplane's performance.

THE SHORT-FIELD LANDING

While it would seem that the shortest short-field landing results from a gliding approach, quite the reverse is true. A powered approach—in which you maintain the desired airspeed with pitch, and control the glide path with power—should be used when you need to extract every last bit of landing performance from the airplane. Small changes in pitch will make the airspeed behave, and you can alter the glide path as required by varying the power setting; more power when you need to fly up, and a little less power when you're too high.

The objective of this operation is to bring the airplane to a complete stop in the shortest possible distance after crossing a 50-foot obstacle at the threshold of the runway (the obstacle is simulated during training). The short-field landing involves the distance through the air while descending from 50 feet, plus the ground roll to get stopped; therefore, the lowest *safe* airspeed on final approach is the key.

Setting up for a short-field landing, fly a normal approach so that you can continue to use the visual clues you've developed throughout your training, but extend the downwind leg a bit, at least fifteen seconds. This provides a longer final approach, gives you more time to set up and adjust the glide path, and forces you to use power as you descend toward the runway. If your airplane is equipped with wing flaps, extend them all the way; this steepens the glide path, allowing you

Maximum Performance Operations

to pass over the obstacle and get down to the runway quickly.

Airspeed should be stabilized early on the final approach at the value specified in your airplane's handbook, and it's important to maintain this airspeed throughout the approach, right down to the runway. (If you can't find a recommended short-field airspeed for your airplane, have your Flight Instructor work out a safe speed; often best-angle-of-climb airspeed is a good place to start.)

Pass over the obstacle, and continue the stabilized glide path right down to the runway; the low airspeed will usually require a pitch attitude in the neighborhood of the normal landing attitude, so very little flare is required. Maximum performance takes on a slightly different meaning at this point, because you are flying close to a stall, you're close to the ground, much judgment is required, and you need to be right the first time—once the airplane is on the ground, you can't change your mind!

Touch down in a normal attitude (on the main wheels for the nosewheelers, three-point for the taildraggers), close the throttle, get the wheel or stick all the way back, and get on the brakes; not enough to skid the tires, but enough to get the airplane stopped as soon as possible. If wing flaps have been used, retract them when you're solidly on the ground; this "dumps" some of the lift and puts more weight on the main wheels, which helps braking action considerably.

As you might have figured out by now, a short-field landing is a rather busy, complicated operation. Recreational Pilots should not expect to produce book results at the outset; practice on a normal-size runway until you learn your personal limits, then add some distance to that. Remember that you can land most airplanes on fields that are too short for takeoff. When a runway looks too short for your airplane, it probably is—stay away from problem situations.

THE SOFT-FIELD TAKEOFF

One of the primary reasons for establishing the Recreational Pilot certificate was to enable people to use the "grass-roots" airports that abound in the United States. "Grass-roots" is a good description, because a lot of these small airfields have no paved runways; the very good ones are surfaced with well-drained, well-maintained sod, and takeoffs and landings are very pleasant experiences.

But there are just as many unpaved runways that are bumpy, muddy, made of gravel, or grass that doesn't see a mower's blade quite often enough. When this is the case, there are concerns about mechanical damage to the airplane's undercarriage and the problems of increased takeoff and landing distances because of the rough surface. A different technique must be employed if the airplane is to get off and on again with no damage, and within the confines of the runway.

The technique for a soft-field takeoff should be used for any unpaved runway, because the objective serves all situations equally well; the idea is to get the airplane into the air at the lowest possible safe airspeed, thereby getting it away from the bumps and perhaps long or wet grass that create a great deal of drag. You'll use ground effect, that magic cushion of air that enables the airplane to fly at a slightly lower airspeed—but *only* for a few feet above the ground.

You should roll onto the runway at the highest taxi speed commensurate with the condition of the surface, so that you already have some airspeed in the bank when you start the takeoff. Add full power as soon as you are straightened out on the centerline, and use as much back pressure on the wheel or stick as is necessary to lift the nosewheel just barely

Maximum Performance Operations

off the ground (you'll be shown the pitch attitude and visual cues to make this happen). In a taildragger, lower the nose very slightly to get the tailwheel off the ground, and then—in both types of airplanes—maintain this attitude until you're airborne; the airspeed will be considerably lower than for a normal takeoff, but that's what using ground effect is all about.

When the airplane is definitely off the ground, lower the nose slightly to maintain level flight, and let the airplane accelerate. Don't forget that the objective of this operation is simply to leave the ground at the lowest possible airspeed; from there on, normal climb procedures take over. As the airspeed needle approaches the normal climb indication, change the pitch attitude and fly away.

THE SOFT-FIELD LANDING

Nearly a mirror image of the soft-field takeoff, a soft-field landing is intended for the same type of unpaved surfaces: rough, wet, or covered with long grass. The objective should be to touch down as slowly and softly as possible, to minimize potential damage to the landing gear, and optimize the landing distance.

Fly the same pattern and approach as for a short-field operation, but as the airplane enters ground effect (about one half of its wingspan above the ground), add just enough power—it doesn't take much!—to lower the airplane gently to the surface in a nose-high attitude. Touchdown should take place on the main wheels with a tricycle-geared airplane, on all three wheels with a taildragger.

Recreational Flying

After touchdown, reduce the power to idle, and hold the nose in the landing attitude as long as you can; this guarantees that the nosewheel (the weakest part of the landing gear system) will contact the surface at a very low speed. In a taildragger, keep the stick *all the way back* after touchdown to prevent a nose-over in mud or very long, wet grass, or deep snow, for that matter.

20

Emergency Procedures

THE MOST LIKELY EMERGENCY for a Recreational Pilot is the obvious one—engine failure—and the procedures for that unhappy circumstance were thoroughly discussed in Chapter 11. But to pretend that nothing else could go wrong with an airplane or its pilot would be leading you down the garden path; you need to prepare yourself for a couple of situations—fires and low-visibility encounters—that have the potential to cause real problems.

AIRCRAFT FIRES

A fire on the ground is usually the result of overpriming an aircraft engine in an attempt to get it started, and it usually occurs during cold weather, when more priming is required. Unfortunately, the fire will probably start in the carburetor, which you can't see from the cockpit, and your first indication will be smoke or smell. (It's not a bad idea to have someone stand outside during a cold-weather start to let you know if all is not well.)

Regardless, when you suspect an engine fire on the ground, *continue cranking* with the starter to draw the fire into the engine where it can burn harmlessly, and pull the mixture control out all the way to shut off the fuel supply. If the fire doesn't go out right away, *get out of the airplane* and get away from it, for obvious reasons. Don't fly the airplane until it's inspected; there can be significant damage, which is detectable only by an aviation mechanic.

Fires in the air are something else. There are two major sources of airborne fires: fuel and electricity. In almost every case, remove the source and you've taken care of the fire. When a fire in the engine compartment is being fueled by gasoline, take the obvious step—shut off the fuel supply, first by means of the mixture control (that's the quickest way), then with the fuel selector (that's the permanent way).

Of course either of those actions renders the airplane powerless in short order, so you must set up immediately for a forced landing. If the fire doesn't go out and you have plenty of altitude, you might consider a dive to extinguish the flames, but don't give up more altitude than you can spare. Getting the airplane safely on the ground as soon as possible should be your number-one objective at this point.

LOW-VISIBILITY ENCOUNTERS

Let's put this in perspective right away: there is no excuse for a Recreational Pilot trying to fly in visibilities so low that

Emergency Procedures

all his visual cues are lost. The 3-mile visibility limit is the most important one in your book of rules, and for this reason: neither you nor your airplane is equipped for instrument flying. It is not necessary to explain this in detail (you'll study human and equipment limitations if you go on to the Private Pilot program); for now, suffice it to say that *a pilot cannot survive when deprived of outside visual cues unless there are adequate flight instruments available, and the pilot has been trained to interpret their indications.* It's not a matter of individual skill or bravado or reflexes, it's a physiological fact—it simply can't be done!

Because it's so vital to remain in good visibility ("good" is when you can see enough of the horizon to keep yourself oriented . . . and three miles is barely enough), you must stay well clear of precipitation and clouds, you must be on the ground well before darkness takes over, and you must recognize the signs of impending low visibility in your flying area.

If you should wander into a situation where ground features are beginning to disappear, turn around, slow down, descend, do whatever you have to do in order to keep that all-important horizon in sight. If that means flying at a much lower altitude than normal, remember that there are towers, power lines, any number of obstacles out there; you must be more alert than ever before.

But most important of all, you must *land* the airplane rather than enter the clouds. Even if it means landing somewhere other than an airport, even to the extent of possibly doing damage to the airplane, getting down under control is a much better option than what will happen if you attempt to continue flight in low- or no-visibility conditions.

Remember: *you can't handle it!*

OTHER EMERGENCIES

Suppose something inside the airplane catches fire, or the cabin door comes open in flight, or a bird shatters the windshield, or any one of a thousand other circumstances that might befall a pilot?

There's no way you can make specific preparations for anything that might come along, but there is a basic philosophy that you can adopt that will give you the best chance of coming out on top. That is to firmly resolve that no matter what, you will *continue to fly the airplane*. Do what you think is best for the situation at hand, but fly the airplane, as long as there's an airplane to fly.

If you lose control, nothing else matters.

21

Safety of Flight Procedures

TALK ABOUT your worn-out clichés, how about this one? "If God had meant man to fly, he'd had given him wings." And I'll bet that someone brought that to your attention when you let it be known that you were going to learn to fly; there are a lot of people who consider being aloft the most dangerous thing they could do.

But such is certainly not the case. To be sure, there is some additional risk when you compare flying to watching television or eating dinner, but you're *always* at risk whenever you move . . . and therein lies the difference. Flying adds height to the risk equation, but removes the frequency of encounters with others; but when altitude is managed sensibly and pilots take every precaution to avoid conflicts, flying can be as safe as it is enjoyable.

Most small-airplane accidents involve the pilot's attempt

to continue flying in weather for which he is not trained or for which the airplane is not equipped, intentionally flying too close to the ground, or lack of vigilance to see and avoid other aircraft. All of these can be prevented, and for the Recreational Pilot, that statement can be printed in capital letters: ALL OF THESE CAN BE PREVENTED.

I can guarantee that you won't prevent incidents or accidents in the air just by hoping that they won't happen to you; there are procedures and techniques that must be practiced and used every time you fly, and knowledge that must be refreshed and expanded at every opportunity. That's what this chapter is all about.

PHYSICAL AND PSYCHOLOGICAL CONSIDERATIONS

During the beginning stages of your flight experience, you will need most of your aeronautical skill and knowledge to fly in normal situations, and you should count on using whatever's left—the extra skill—in abnormal circumstances such as strong winds, short or soft fields, emergencies, and so on.

To be sure that the extra "edge" is there when you need it, don't fly when you're ill, taking medication, or just plain tired out. In addition to the obvious safety aspect, you'll be pleasantly surprised to discover how much better you perform as a pilot, and how much more you'll enjoy your flying, when you feel good.

Safety of flight has more to do with pilots' mental attitudes than most of us realize; it deserves your undivided attention from the time you leave the ramp until the airplane is tied down again at the end of the flight. Don't fly when you're worried or preoccupied with other things, or when you feel that for whatever reason you *must* fly today, regardless of circumstances. That's usually the first link in the chain of events that leads to an accident. When you sense that an upcoming flight operation (short field, strong crosswind, threatening weather, whatever) may be beyond your skill level, turn away from it. You can always come back for another try later. Beware of bravado and "get-home-itis," a disease that's not unique to aviation, but which runs rampant through the pilot population. If "it" doesn't look right, or doesn't feel right, it's probably unsafe for you today . . . listen to that inner voice.

COLLISION AVOIDANCE

Recreational Pilots (and everybody else at uncontrolled airports!) operate in an environment of self-control, unlike a tower-equipped facility where you are assigned a position in the traffic pattern, cleared for landing and takeoff, and generally looked after by someone else. This self-control system breaks down unless every pilot knows, understands, and observes the rules that govern aircraft movement at an uncontrolled airport. The first step in collision avoidance at the airport, then, is to *always* fly the pattern as it should be flown; that way, you'll be where other pilots are looking for you, and vice versa.

Understand your visual limitations. One of the most important is that your eyes tend to focus on a close-in point when there's nothing out there to look at. So you must keep your eyes moving if you expect to see other airplanes; look from one side to the other and up and down with a regular, frequent scan of all the airspace within your field of vision. The FAA and others have published some very good material on the subject of visual scanning techniques; they are highly recommended.

When you spot another aircraft that might represent a threat to your safety, get out of the way. Of course the procedures in the regulations must be followed, but the most important part of collision avoidance is *getting out of the way!* You can always argue about the way you did it later. Keep the other airplane in sight during your avoidance maneuver.

Sometimes another airplane is close enough when you spot it that there's no doubt about the threat; but frequently you'll see someone else so far away that you can't really tell whether your flight paths will cross or not. In this situation, relative movement is the key; as soon as you see a distant aircraft, notice its position on your windshield or side window, then look back in a few seconds to see if the relative position has changed. If the intruder has moved at all, you're not going to meet; but if the other airplane stays in the same place (no relative movement), do something *right now*. Climb, descend, turn right or left enough to get some relative movement started. You must change your flight path to avoid a collision.

AVOIDING WAKE TURBULENCE

No matter how streamlined an airplane may be, it displaces air as it moves along, and since any natural system returns to equilibrium just as soon as it can, the air will try to get back where it was before the airplane showed up. When *your*

Safety of Flight Procedures

airplane gets in the way as the disturbed air tries to return, there's going to be a bump or two—wake turbulence. Common sense dictates that you avoid the area immediately behind another aircraft, particularly one that is much larger than yours.

There's another part to the wake turbulence problem, and it's known as *wingtip vortices*. The higher-pressure air under the wing, the very essence of lift, spills up and over the wingtips as the airplane moves forward, and the result is a pair of rapidly rotating horizontal vortices that trail behind. They tend to level off about 500 feet below the airplane's flight path, where they may persist for several minutes.

If you fly *across* such a disturbance, you will experience a sharp updraft, then a strong downdraft, and finally another updraft as you exit the wake. If you encounter the vortices while flying in the same direction as the generating airplane, you'll be rolled to the right or the left, depending on which vortex you encounter.

Once again, the strength of this aerodynamic phenomenon is directly related to the size of the generating airplane, and the ability to cope with wingtip vortices depends on the relative size of your airplane. Rather than try to figure out how large an airplane you can safely follow, play it safe and avoid that area immediately below and behind *all* airplanes larger than yours. When there are larger airplanes (or helicopters, for that matter), plan your takeoffs, patterns, and landing approaches so that you are never in the hazardous area below and behind.

INDEX

Accelerated stalls, 101
Accidents, 145, 149, 151–53, 223–24
  general flight rules and, 17–18
  Instrument vs. Recreational Pilots and, 113, 221
  reporting, 5
  traffic patterns and, 114
  See also Safety
Aerobatics, 18–19, 57, 103
Aerodynamic lever, 195, 196
After-landing checklist, 119–20
Ailerons, 22, 25, 27, 41–42
  before-takeoff check of, 89
  crosswind and, on ground, 87
  in crosswind landing, 136, 140–41, 144
  in crosswind takeoff, 137–39, 144
  in turns, 66, 67
Air density, 51
  density altitude, 201–203
  descent and, 64
  fuel mixture strength and, 75
Airfow in production of lift, 34–37
Airline Transport Pilot Certificate, 7
Airman Certification Regulations, 92
Airman's Information Manual (U.S. Government), xv

Air masses, 176–81
Airport operations, 145–52
  Air Traffic Control, 11, 145, 153, 155
  controlled vs. uncontrolled, 145, 155–56
  inbound procedures, 150–51
  outbound procedures, 148–50
  right-of-way rules, 17–18, 113, 148, 151
  runway selection, 147–48
  spectator hazards, 151–52
  See also Traffic pattern procedures
Airport Radar Service Area (ARSA), 156
Airport Traffic Area, 155
Air pressure, 43
  altimeters and, 52–54
  in lift production, 34–37
  performance and, 201–203
  speed of descent and, 63
  speed measurement and, 49–51
  See also Fronts, weather
Airships, *see* Blimps
Airspace restrictions, 153–59
  general, 154
  special-use, 156–58
  temporary, 158–59
  on type of airport, 113, 145, 153, 155–56

Index

Airspace restrictions (*cont.*)
  weather conditions and, 10, 11, 19–20, 153, 154, 168–69, 182, 186–89, 221
Airspeed:
  in accelerated stall, 101–102
  for climbing, 60–62, 125, 206–209, 213
  in dead reckoning, 165
  decreasing, 40–41
  for descent, 63–65
  excessive, on landing, 143
  and lift, 36
  limitations on, 16, 63–64
  minimum controllable (MCA), 94–96
  normal cruise, 92–93
  for short-field landing, 215
  in slow flight, 93–94
  in straight and level flight, 58, 92–93
  trim tab and, 43
Airspeed indicator, 16, 48–51, 89
Air Traffic Control, 11, 145, 153, 155
Airworthiness, regulation check on, 88
Airworthiness Certificate, 14–15
Alcohol, 13–14
Alert Areas, 158
Altimeter, 16, 48–49, 52–54, 89
  straight and level flight and, 58–59
"Altimeter setting," 54
Altitude:
  clouds and, 188–89
  engine failure and, 208–209 (*and see* Engine-out landings)
  general limits on, 154, 187
  for ground reference maneuvers, 103–104
  mixture control and, 75
  regulations on, 10, 11, 19, 130
  traffic patterns and, 112–14, 150, 151
  in training stalls, 97
  turning and, 67
Angle of attack, 35–37, 40
  propeller effects and, 45–46
  rate-of-climb airspeed and, 61
  and stall, 96, 97, 102, 185
  turning and, 67
Angle of bank, *see* Bank
Approach:
  adjusting, for wind, 121–22
  in crosswind, 136–37, 139–40
  engine-out, 125–29
  go-around on, 108
  normal, 115–18
  to strange airport, 150–51
  in traffic pattern, 114
  See also Landing
Approach-landing stall series, 100
Atmospheric pressure, *see* Air pressure
Attitude, *see* Pitch attitude
Aviation gasoline (avgas), 72, 77, 80
  See also Fuel
Axes of rotation, 26

Balloons, 18
Bank, 43, 59, 60
  in crosswind landing, 136, 139–42
  in training stalls, 98–102
  for turns, 65–69
  in S-turns, 105–106
  See also Roll
Battery, storage, 76
Before-takeoff procedure, 88–91, 148–50
Best angle of climb airspeed (V_x), 61, 207–208
Best glide airspeed, 125
Best rate of climb airspeed (V_y), 61, 125, 206–209, 213
Biplanes, 24
Blimps, 18, 134
Bonded surfaces, 24

Index

Brakes, 30
  checking, 85
  in taxiing, 85–86
"Break," 98
"Burbling," 37

Cantilever bracing, 25
Carburetor, 74, 90–91, 116, 125
Center of gravity (CG):
  crosswind and, 137, 143
  load placement and, 194–99
Center of lift (CL), 194
Centrifugal force, 67–68
  in accelerated stall, 102
  crosswind and, 87, 137
  load factor and, 191, 192
Certificated Flight Instructor
  (CFI), see Flight Instructor
Certification:
  FAR on, 5–12
  medical, 7–8
  training operations and, 92
  types of, 6–7
  See also Recreational Pilot Certificate
Charts, aeronautical, 154, 161–62
CIFFTRS before-takeoff checklist, 88–91
Cirrus clouds, 170
Climbing, 57, 60–63, 125
  performance and, 200, 206–209
  in short-field takeoff, 213
Clouds, 113, 153, 169–72, 175, 177, 178, 180, 181, 188–89, 221
Cold fronts, 176–80, 182, 183
Collisions, see Accidents
Commercial Pilot Certificate, 6–7
Compass, magnetic, 16, 48, 89, 164
Components, airplane, 21–30
  diagram of, 22
  See also specific parts
Construction materials, airplane, 21–23
Control surfaces, 22, 25–27
  before-takeoff check of, 89
  preflight check of, 80
  primary vs. secondary, 25, 27, 43
  See also Ailerons; Elevator; Rudder; Trim tabs
Control Zone, 155
Cooling system, 71–72
Course:
  in dead reckoning, 165
  definition of, 104
Crosswind, 134–44, 174
  airplane limits and, 142
  dead reckoning and, 164–65
  landing in, 134–37, 139–44
  takeoff in, 135, 137–39, 144
  taxiing in, 87, 137, 141
Cruise:
  engine failure in, 125, 128–30
  normal, 92–93
  See also Straight and level flight
Cumulus clouds, 170–72, 178, 180, 181
Cylinders, 71–72, 76

Datum, 196, 197
Dead reckoning, 163–65
Density altitude, 201–203
Descending, 57, 63–65
  See also Approach; Landing
Detonation, fuel, 76–77
Dew point, 169–72, 181
Disc brakes, 30
Drag, 31, 37
  definition of, 32, 33
Drift correction, 104–106, 112
  crosswind and, 136, 138–42, 144
Drugs, illegal:
  transportation of, 14
  use of, 13–14
Dry fronts, 178

Effective weight, 192
Electrical systems, 76
  and fires, 220

231

Index

Elevator, 22, 25–27, 39–40, 43–44
  before-takeoff check of, 89
  placing of load and, 194–96
  in short-field takeoff, 213
  See also Trim tab
Emergencies, 219–22
  fires, 219–20
  regulations and, 13
  See also Accidents; Engine-out landings
Emergency locator transmitter (ELT), 16
Empennage, 27
Empty weight, 193–94
Engine, 27–28, 70–73
  runup before takeoff, 90–91
  See also Powerplant
Engine controls, 43
  See also Mixture control; Throttle
Engine-out landings, 123–31
  from cruise, 125, 128–30
  on takeoff, 125–27
Engine-starting techniques, 81–83

Federal Aviation Administration (FAA), 3–5, 13
  and airspace restrictions, 153, 154
  and flight reviews, 11–12
  Flight Service Stations (FSS) of, 139, 157, 159, 165, 185–86
  physiological training courses of, 154
  RPC introduced by, ix–x
  on visual scanning, 226
Federal Aviation Regulations (FARs), 5–20, 190
  general flight rules, 17–20
  general operating rules, 12–17, 48, 55, 74, 76, 77, 88, 190
  on preflight procedures, 78
  for Recreational Pilots, 10–12
  for Student Pilots, 7–10

Temporary Flight Restrictions, 158–59
  types of certification, 6–7
Fiberglass in airplane construction, 23, 24
Fires, 219–20
Flaps, *see* Wing flaps
Flight controls, *see* Control surfaces; Engine controls
Flight Guide, 15
Flight Instructor (CFI), xv, 8–9, 64, 79, 81–82, 97, 102, 116, 124, 131, 141, 160, 166, 167, 175, 188, 206, 212, 215
  certification of, 7
Flight reviews, 11–12
Flight Service Stations (FSS), 139, 157, 159, 165, 185–86
Flight Standards District Offices, 154
Flight Training Handbook (U.S. Government), xv
Fog, 171, 175–77, 189
Formation flying, 17
Four-cycle engine, 72–73
Fronts, weather, 174–83
Fuel:
  before-takeoff check of, 89
  engine failure and, 125
  and fires, 220
  preflight check of, 14, 80
  pre-landing check of, 115
  in priming process, 81–82
Fuel-grade limitations, 76–77
Fuel quantity gauge, 16, 74, 89
Fuel selector, 74
Fuel system, 72–75
Full-monocoque design, 24
Fuselage, 22–24

Gasoline, *see* Aviation gasoline; Fuel
G force, 102, 191, 192
  See also Centrifugal force
Gliders, 16, 18

232

Index

Go-around, 108, 142
Gravity, *see* G force; Weight
Ground loop, 143
Ground reference maneuvers, 103–108
   rectangular pattern, 104–105, 112
   straight-line tracking, 104
   S-turns, 105–106
   turns about a point, 106–108
Ground-steering system, 85
Ground surface, taxiing and, 85–88
GUMPS pre-landing check, 115
Gyroscopic effect, 47

Hail, 172, 183–84
Hand-propping, 83
Heading, definition of, 104
Helicopters, 18, 227

Icing, carburetor, 90–91, 116, 125
Instrumentation, 48–56
   before-takeoff check of, 89
   minimum requirements on, 16, 48–49, 55, 73–74
   See also specific instruments
Instrument pilots, 113, 153, 155

Jet engines, 28, 45

Landing:
   adjusting, for wind, 121–22
   airplane weight and, 203–204
   airport procedures for, 150–51
   crosswind, 134–37, 139–44
   emergency, *see* Engine-out landings
   no-flap, 123–24
   normal, 115–20
   performance and, 203–206, 211, 214–18
   preparation for, 115
   short-field, 214–16

soft-field, 217–18
wheel, 132–33, 143–44
See also Approach
Landing gear, 22, 29–30, 33
   pre-landing check of, 115
   retractable, 11, 24, 29, 115
Landing site, emergency, 125–31
Leaning, 75, 82
Level flight, *see* Straight and level flight
Leveling off:
   from climb, 62–63
   from descent, 64–65
Lift, 28, 31–38
   definition of, 31–33
   and descent, 63
   load and, 191, 194
   production of, 33–37, 50–51
   of propeller, *see* Thrust
   in spin, 103
   in turning, 65–67
   vs. weight, 37–38
Loading, 190–99
   load factor, 191, 192
   placing of load, 194–99
   performance and, 191, 203–204
   restrictions on, 16, 190, 194
Logbook, 12
Lubrication system, 73–74

Magnetic compass, 16, 48, 89, 164
Magnetos, 76
Magneto switches, 79
Manifold pressure:
   limitations on, 77
   normal cruise speed and, 93
Manifold pressure gauge, 16, 48–49, 55–56
Maximum allowable gross weight, 193–94
Medical certification, 7–8
Military Operations Areas (MOAs), 157–58
Minimum controllable airspeed (MCA), 94–96

Index

Mixture control, 72–73, 74–75, 79, 89, 115
   in fires, 220
Moment, 196–98
Monoplanes, 24
Multiplanes, 24

Nacelle, 28
National Transportation Safety Board, 5
Navigation, aerial, 160–67
   charts and, 154, 161–62
   dead reckoning, 163–65
   orientation, 166–67
   pilotage, 162–63
Never-exceed speed (V_{ne}), 63–64
Normal cruise, 92–93
Notices to Airmen (NOTAMs), 159

Oil pressure gauge, 16, 73–74
Oil system, 73–74
Oil temperature gauge, 16, 73–74
Orientation procedures, 166–67
Owner's Manual, 15, 142
Oxygen pressure, 154

Parachutes, 103
Parasol wings, 25
Passengers, regulations on, 10–12
Performance, 200–218
   density altitude and, 201–203
   normal vs. maximum, 200, 211–12
   pilot technique and, 209–10
   runway and, 200, 204–206, 212–18
   vertical, 200, 206–209, 212–14, 216–17
   weight and, 191, 203–204
   wind and, 204–205
P force, 93
   climbing and, 62–63
   crosswind and, 138
   descending and, 64
   in normal takeoff, 110, 111

Pilotage, 162–63
Pilot's Handbook of Aeronautical Knowledge (U.S. Government), xv
Pilot's Operating Handbook, xiv, 15, 79, 142, 212
Pistons, 72, 73
Pitch, 26, 38, 39, 43
Pitch attitude:
   airspeed and, 51, 62
   for climbing, 60–62
   in crosswind takeoff, 138
   in descent, 64–65
   after engine failure, 125
   for normal takeoff, 111
   in roundout and touchdown, 117–19
   in short-field takeoff, 213
   in soft-field takeoff, 216–17
   in straight and level flight, 58–59
   in training stalls, 97–100
   in turns, 69
Pitot tube, 49–51
Plastic in airplane construction, 23, 24
Plywood, molded, 24
Power, thrust and, 43, 70, 72–73
Powerplant, 49, 70–77
   basic features of, 71–73
   electrical systems and, 76
   limitations on, 15, 70–71, 76–77
   oil and fuel systems and, 73–75
   weight of, 71–72
   See also Engine; Propellers
Power setting:
   for climbing, 60
   for descent, 63
   for leveling off, 62, 64–65
   for straight and level flight, 58, 93
   in training stalls, 98, 101
   See also RPM
Precipitation, 172–73, 177–79, 182, 189
Preflight briefing, 139

Index

Preflight inspection, 72, 74, 78–81, 137
Priming, 81–82
Private Pilot Certificate, x, xi, 6, 169, 221
Prohibited Areas, 156–57
Propellers, 28
   in descent, 64
   effects of, 45–47
   fixed-pitch vs. constant-speed, 55–56
   hand-propping, 83
   pre-landing check of, 115
   safety and, 28, 83, 151–53
   in straight and level flight, 59
   and thrust, 28, 32–33, 46–47, 70

Radial powerplants, 28
Radio communications, 145, 150, 152, 153, 156
Rain, 172–73, 178–79, 183, 189
   See also Precipitation
Ratings, pilot, 6
Reciprocating engines, 71
Recreational Pilot Certificate (RPC), ix–x, xiii–xv, 188
   FARs and, 5, 7–12
Rectangular pattern, 104–105, 112
Registration certificate, 15
Regulations/regs, *see* Rules and regulations
Relative humidity, 169
Restricted Areas, 157
Right-of-way rules, 17–18, 113, 148, 151
Roads, landing on, 130
Roll, 26
   ailerons and, 27, 43
   propeller effects and, 45, 47
   in straight and level flight, 59, 60
   takeoff, 51
   See also Bank
Rotation, axes of, 26
Rotation of nose, 213

Roundout, 117–18, 140
RPM (revolutions per minute), 55, 56
   limitations on, 77
   mixture control and, 75
   and normal cruise airspeed, 93
   in runup, 90
   See also Power setting
Rudder, 22, 25–27, 41
   before-takeoff check of, 89
   in crosswind landing, 136, 139–44
   in crosswind takeoff, 137–39, 144
   in normal takeoff, 110–12
   in straight and level flight, 59, 60
   in training stalls, 98–101
   See also Yaw
Rudder trim, 89–90
Rules and regulations, xiv, 3–20, 92, 130
   on accident reporting, 5
   on aerobatics, 103
   FAA and, 3–5, 153
   pilot's responsibility and, 4–5, 13, 20, 190
   right-of-way, 17–18, 113, 148, 151
   for Student Pilots, 7–10
   See also Airport operations; Airspace restrictions; Federal Aviation Regulations (FARs); Visibility
Runup, 90–91
Runup pad, 148
Runway:
   leaving, in normal landing, 119
   selecting, 147–48, 150
   short, 61, 212–16
   soft, 216–18

Safety, 223–27
   airworthiness certificate, 14–15
   collision avoidance, 225–26

Index

Safety (*cont.*)
  general operating rules on, 13–15
  pilot's condition and, 224–25
  propellers and, 28, 83, 151–52
  wake turbulence, 226–27
  See also Accidents
Safety belts, 13, 16, 91, 115
Sea level, altimeters and, 52–54
Sectional Aeronautical Charts, 154, 161
See-and-avoid rule, 17–18
Semimonocoque fuselage, 23–24
Shock absorbers, 29–30
Short-field landing, 214–16
Short-field takeoff, 61, 212–14
Shoulder harnesses, 13, 16
Slipstream effects, 45, 46
Slow flight, 93–94
Snow, 172–73, 178–79
Soft-field landing, 217–18
Soft-field takeoff, 216–17
Solo flights, 8–9
Spark plugs, 76, 81
Special Conservation Areas, 158
Special-Use Areas, 156–58
Spins, 98, 101–103
Squall line thunderstorms, 184
Stabilizers, 22, 27
  horizontal, 22, 27, 38–39
  vertical, 22, 26, 27, 41, 93
Stall(s), aerodynamic, 36–37, 40–41, 92, 96–102
  accelerated, 101–102
  approach-landing series, 100, 185
  MCA and, 94–96
  power-off, straight ahead, 97–98
  power-off, turning, 98–99
  power-on, straight ahead, 98
  power-on, turning, 99–100
  slow flight and, 93–94
  takeoff-departure series, 101
  in touchdown, 117–18
Standard situation, defined, 202

Starting engine, 81–83
Stationary fronts, 180–81
Storage battery, 76
Straight and level flight, 31–32, 57–60
  normal cruise, 92–93
Straight-line tracking, 104
Stratus clouds, 170–72, 175, 180
Struts, external, 25
Student Pilot regulations, 7–10
S-turn, 105–106
Sump, 73

Tachometer, 16, 48–49, 55, 75
Tailwheel/"taildragger" configuration, 29
  normal landing with, 118–119
  normal takeoff with, 111
  short-field takeoff with, 213
  soft-field landing with, 217, 218
  soft-field takeoff with, 217
  wheel landings with, 132–33, 143–44
Takeoff:
  airport procedures for, 147–50
  adjusting, for wind, 120–21
  crosswind, 135, 137–39, 144
  engine failure on, 125–27
  normal, 109–12
  performance and, 203–206, 212–14, 216–17
  short-field, 61, 212–14
  soft-field, 216–17
Takeoff-departure stall series, 101
Taxiing:
  in crosswind, 87, 137, 141
  after landing, 119–20
  before takeoff, 84–88, 148, 216
Temporary Flight Restrictions, 158–59
Terminal Control Area (TCA), 156
Terminal Radar Service Area (TRSA), 156
Tetra-ethyl lead, 77
Throttle, 73–75, 79

Index

in starting, 81, 82
in taxiing, 85
Thrust, 28, 31, 37–40, 43
   asymmetric, 47
   in climbing, 60, 62
   definition of, 32–33
   in descent, 63, 64
   engine instruments and, 55–56
   power and, 43, 70, 72–73
   propellers and, 28, 32–33, 46–47, 70
Thunderstorms, 174–75, 181–84
Torque effect, 47
Tornadic tubes, 184
Touchdown, 117–19
   in crosswind, 136, 140–41
   short-field, 215
   soft-field, 217
Track, definition of, 104
Traffic pattern procedures, 92, 112–14, 145–52, 225
   adjusting, for wind, 120–22
   recommended (diagram), 146
   See also Airport operations; Ground reference maneuvers
Tricycle-gear configuration, 29
   normal landing with, 119
   normal takeoff with, 111
   soft-field touchdown with, 217
   wheelbarrowing with, 143
Trim tab, 22, 27, 43–44, 59
   before-takeoff check of, 89–90
Truss fuselage, 23
Turning, 57, 65–69
   on ground, 85–86
   about a point, 106–108
   S-turns, 105–106
   training stalls in, 98–100
TV weather programs, 187

Undercarriage, *see* Landing gear
U.S. government:
   publications of, xv
   See also Federal Aviation Administration (FAA)

Useful load, 193

V_{ne} (never-exceed speed), 63–64
V_x (best angle of climb airspeed), 61, 207–208
V_y (best rate of climb airspeed), 61, 125, 206–209, 213
Variation, 164
Visibility, 220–21
   fog, 171, 175–77, 189
   Visual Flight Rules (VFR) minimums and, 10, 11, 16, 19–20, 153, 154, 182, 186–88, 221
Visual scanning, 226

Wake turbulence, 226–27
Warm fronts, 176–77, 179–80
Water landings, 131
Weather, 168–89
   and altimeter readings, 53
   clouds, 113, 153, 169–72, 175, 177, 178, 180, 181, 188–89, 221
   fog, 171, 175–77
   fronts, 174–81
   personal observation of, 188–89
   precipitation, 172–73, 177–79, 182, 189
   preflight briefing on, 139
   reports, 139, 185–87
   thunderstorms, 174–75, 181–84
   See also Visibility; Wind
Weathervane effect, 86
Weight, 31, 37–38
   centrifugal force and, 67–68
   definition of, 32, 33
   and descent, 63
   performance and, 191, 203–204
   of powerplant, 71–72
   See also Loading
Wheelbarrowing, 143
Wheel landings, 132–33, 143–44
Wind:
   adjusting approach for, 120–22
   in dead reckoning, 164–65

237

Index

Wind (*cont.*)
  determining direction of, 128, 137, 147, 175
  engine-out landings and, 127–29
  meteorology of, 173–75, 177, 179, 180, 184–85
  and performance, 204–205
  taxiing and, 86–87, 119
  traffic patterns and, 113–14
  See also Crosswind; Ground reference maneuvers
Wind correction, *see* Drift correction
Wind direction indicators, 137, 147
Wing flaps, 115
  adjusting landing for wind and, 122
  before-takeoff check of, 89
  in enginge-out landings, 126–27
  landing without, 123–24
  at MCA, 96
  in recovery from stall, 97
  in short-field landing, 214
  in short-field takeoff, 212

Wings:
  ailerons and, 41–43
  construction of, 22, 24–25
  in crosswind landing, 136, 139–42
  horizontal stabilizer and, 38–39
  and lift, 34–37, 51–52
Wingtip vortices, 227
Wright brothers, 22, 48, 49, 86

Yaw, 26, 41, 43, 59–60, 93
  during climb, 62
  in crosswind takeoff, 138
  landing and, 119
  in leveling off, 64–65
  at MCA, 95
  in normal cruise, 93
  in normal takeoff, 110, 111
  propeller effects and, 45–47, 59
  in slow flight, 94
  in training stalls, 98–102
  in turns, 68